VOLUME EIGHTY NINE

ADVANCES IN
COMPUTERS

VOLUME EIGHTY NINE

ADVANCES IN
COMPUTERS

Edited by

ATIF MEMON
University of Maryland
4115 A.V. Williams Building
College Park, MD 20742
Email: atif@cs.umd.edu

Amsterdam • Boston • Heidelberg • London
New York • Oxford • Paris • San Diego
San Francisco • Singapore • Sydney • Tokyo
Academic Press is an imprint of Elsevier

Academic Press is an imprint of Elsevier
The Boulevard, Langford Lane, Kidlington, Oxford, OX51GB, UK
32, Jamestown Road, London NW1 7BY, UK
Radarweg 29, PO Box 211, 1000 AE Amsterdam, The Netherlands
225 Wyman Street, Waltham, MA 02451, USA
525 B Street, Suite 1800, San Diego, CA 92101-4495, USA

First edition 2013

Library of Congress Cataloging-in-Publication Data
A catalog record for this book is available from the Library of Congress

British Library Cataloguing-in-Publication Data
A catalogue record for this book is available from the British Library

ISBN: 978-0-12-408094-2
ISSN: 0065-2458

For information on all Academic Press publications
visit our website at *store.elsevier.com*

Printed and bound by CPI Group (UK) Ltd, Croydon, CR0 4YY

CONTENTS

3. Model Inference and Testing
Muhammad Naeem Irfan, Catherine Oriat, and Roland Groz

4. Testing of Configurable Systems
Xiao Qu

5. Test Cost-Effectiveness and Defect Density: A Case Study on the Android Platform
Vahid Garousi, Riley Kotchorek, and Michael Smith

PREFACE

This volume of Advances in Computers is the 89th in this series. This series, which has been continuously published since 1960, presents in each volume five to seven chapters describing new developments in software, hardware, or uses of computers.

The computing landscape is ever changing. Today, we increasingly talk about engineering mobile apps that execute on our phones and tablet devices. We engineer our software with increasing focus on its future maintenance. Because of our end-users' diverse needs, our software systems are highly configurable. All these trends call for changes to our existing software engineering processes.

In this volume, we focus on one important software engineering activity, namely software testing. This volume is a compilation of a set of five chapters that study issues of software testing of mobile applications, evolving software, and highly-configurable software. The authors of these chapters are world leaders in their fields of expertise. Together these chapters present, for certain key domains, the state-of-the-art in software testing.

The rise in popularity of mobile applications for mobile devices and the growth estimates for this market make mobile application development a strategic business sector. As a variety of new scenarios for mobile devices and applications emerge, users and developers require improved reliability, usability, performance, and security. In such a context, open platforms for mobile application development, such as the Android operating system, are assuming a preponderant role. To satisfy this growing request for high quality applications, developers must devote greater effort and attention to software development processes. In particular, testing and its automation play a strategic part for assuring the quality of applications. In Chapter 1, "Testing Android Mobile Applications: Challenges, Strategies, and Approaches," by Domenico Amalfitano, Anna Rita Fasolino, Porfirio Tramontana, and Bryan Robbins, the authors analyze the main challenges and open issues in the field of mobile application testing for the Android platform, with an emphasis on advances in the field.

Software changes, such as bug fixes or feature additions, can introduce software bugs and reduce code quality. As a result, tests that passed earlier may no longer pass, thereby exposing a regression in software behavior. Chapter 2, "Regression Testing of Evolving Programs," by Marcel Böhme, Abhik Roychoudhury, and Bruno C.d.S. Oliveira surveys recent advances in determining the impact of the code changes onto the program's behavior and other syntactic program artifacts.

Software models may be used to steer testing and model checking of software systems. Such models may be obtained from behavioral traces, available specifications, knowledge of experts, and other such sources. Model inference techniques typically extract structural and design information of a software system. In Chapter 3, "Model Inference and Testing," by Muhammad Naeem Irfan, Catherine Oriat, and Roland Groz, the authors discuss passive and active model inference techniques. They present an approach that switches between model inference and testing. In the model inference phase it asks membership queries and records answers to obtain a model of a software system under inference. In the testing phase it compares the obtained model with the system under inference. If a test for a model fails, a counterexample is provided which helps to improve the model.

Configurable software system allows users to customize its applications in various ways, and is becoming increasingly prevalent. Testing configurable software requires extra effort over testing traditional software because there is evidence that running the same test case under different configurations may detect different faults. Differentiating test cases and configurations as two independent factors for testing, one must consider not just which test case to utilize, but also which configurations to test. Ideally, an exhaustive testing approach would combine every test case with every possible configuration. But because the full configuration space of most software systems is huge, it is infeasible to test all possible configurations with all test cases. Instead, selection techniques are necessary to select configurations for testing a software system, and to select test cases for the different configurations under test. Despite successful selection techniques, sometimes it is still costly to run only selected configurations and test cases. In particular, the cost is magnified when new features and functionality are added as a system evolves, and the new version is regression tested. Regression testing is an important but expensive way to build confidence that software changes do not introduce new faults as the software evolves, and many efforts have been made to improve its performance given limited resources. Test case prioritization has been extensively researched to determine which test cases should be run first, but has rarely been considered for configurations. In Chapter 4, "Testing of Configurable Systems," Xiao Qu introduces issues relevant to testing configurable software systems and presents techniques for both selection and prioritization of these systems.

In Chapter 5, "Test Cost-Effectiveness and Defect Density: A Case Study on the Android Platform," Vahid Garousi, Riley Kotchorek, Michael Smith assess test coverage, fault detection effectiveness, test cost-effectiveness, and defect density in code-base of version 2.1 of the Android platform. Their results will help readers to get a better view on test coverage, fault detection effectiveness, test cost-effectiveness, and defect density in the Android code-base.

I hope that you find these articles of interest. If you have any suggestions on topics for future chapters, or if you wish to be considered as an author for a chapter, I can be reached at atif@cs.umd.edu.

Atif M. Memon
College Park, MD, USA

CHAPTER ONE

Testing Android Mobile Applications: Challenges, Strategies, and Approaches

Domenico Amalfitano*, Anna Rita Fasolino*, Porfirio Tramontana*, and Bryan Robbins†
*Dipartimento di Informatica e Sistemistica, Università di Napoli Federico II, Via Claudio 21, 80125 Napoli, Italy
†Department of Computer Science, University of Maryland—College Park, College Park, Maryland, USA

Contents

Advances in Computers, Volume 89
ISSN 0065-2458, http://dx.doi.org/10.1016/B978-0-12-408094-2.00001-1

Abstract

Recently, the rise in popularity of mobile applications for mobile devices and the growth estimates for this market make mobile application development a strategic business sector. As a variety of new scenarios for mobile devices and applications emerges, users and developers will require improved reliability, usability, performance, and security. In such a context, open platforms for mobile application development, such as the Android operating system, are assuming a preponderant role.

To satisfy this growing request for high quality applications, developers must devote greater effort and attention to software development processes. In particular, testing and its automation play a strategic part for assuring the quality of applications.

This chapter analyzes the main challenges and open issues in the field of mobile application testing for the Android platform, with an emphasis on advances in the field. We present suitable and effective principles, guidelines, models, techniques, and technologies for Android application testing and conclude with an outline of future perspectives.

1. INTRODUCTION

We define mobile applications as self-contained software designed for a mobile device and performing specific tasks for mobile users. Although the first mobile applications appeared about 10 years ago, their diffusion increased exponentially with the Apple App Store platform opening in July 30th, 2008 [29]. Since that date, the number of third-party applications available from the App Store grew from the initial few hundreds to about half a million applications available on July 30th, 2011, with 15 billions downloads since the store's launch [30].

After the launch of the Apple App Store, several other digital distribution platforms appeared that also provide mobile software to mobile devices. These platforms include both application marketplaces that are native to the major mobile operating systems (such as Android, Palm webOS, BlackBerry OS, Symbian OS, Windows Phone 7, etc.) and third-party platforms offered by commercial organizations such as Amazon (Amazon AppStore), General Software (with Cellmania), and others [79].

With the success of these marketplaces where users of mobile devices can browse and download applications that are available either free or at a cost, the demand for mobile applications continues to grow. Moreover, due to the emerging trend of using remote/mobile applications in and between business and industrial settings, the demand projects to increase even more. Mobile applications indeed provide a means for accessing applications from remote locations and with mobile devices, for collaborating and sharing files with co-workers across the globe, for improved and remotely available productivity tools, and for supporting a growing number of mobile and remote workers.

There are two main types of mobile applications: ones that must be installed on the device (either pre-installed or downloaded from market-places) and mobile Web applications. The former can be further categorized into apps written for a specific type of handset (e.g., iPhone applications, which must be targeted for iPhone) and applications that may run on many handsets, typically written in Java. In contrast, Web applications reside on a server and users access the application over the Internet by means of a Web browser deployed on the mobile device. In this aspect, these applications do not differ very much from traditional Web applications developed for desktop PC, and typically use the same Web technologies, like HTML, CSS, and Java.

Mobile applications providing a data-based service may also be classified based on their software architecture [73]. This classification distinguishes among Client-only applications, Client-Server applications, Content appli-cations, and applications using native APIs. Client-only applications need to be installed on the device and do not have any server-side counterpart in the network. Client-Server apps depend on complementary client and server components that communicate with request-response sequences. Content applications are a particular type of Client-Server apps where the client implements all necessary business logic but fetches content from a server. Some applications using Native APIs can be considered a form of Client-only apps that invoke other native applications like video players, alarm schedules, contact lists, messaging APIs, etc. to perform tasks.

Independent of their type, most mobile applications are usually small and developed by a small team (one or two people) responsible for conception, design, and development [76]. The team usually works on strict timelines and under the pressure of short time-to-market. Teams use powerful development tools and frameworks, but rarely adopt any formal development process. This approach may suit small- or medium-sized applications. However, as mobile applications become more complex and business-critical, use of well-defined software engineering techniques becomes essential. In particular, to assure the necessary quality of these applications, testing activities require greater effort and attention.

In recent years, many research and industrial initiatives have aimed to define effective testing principles, techniques, and tools for mobile applications [34, 39, 66, 67, 70, 73, 77]. Due to the huge growth in popularity of applications for the Android platform in recent months, Android applications also have a need for effective testing techniques, strategies, and tools. The Android operating system continues to rise in popularity, currently listed as the second most popular mobile OS, surpassing BlackBerry and iPhone OS in 2011. That success projects to continue into 2012 [44]. The platform's popularity solidifies the importance of assuring the reliability of Android mobile applications. Assuring reliability will serve as a strategic factor in the sustainability and expansion of that success.

Android application testing represents a challenging activity, with several open issues, specific problems, and questions. As an example, most developers remain largely unfamiliar with the Android development platform, leaving their applications prone to new kinds of bugs. Android applications indeed differ from standard client-server applications as well as traditional event-based desktop GUIs. The structure of Android apps centers instead around specific software components offered by the Android application framework, which require specific management rules and a particular lifecycle [10].

Moreover, due to its novelty, the Android development platform still lacks maturity, as some recent studies conclude [55]. This immaturity further confirms the necessity of accurate testing activities aimed at assuring the quality of these applications.

As with all mobile applications, specific constraints of handheld devices influence the development and testing strategies for Android applications. For example, constraints such as the heterogeneity of mobile device hardware and software configurations, the limited nature of underlying hardware resources, and the scarceness of mechanisms offered by the platforms for preventing security attacks all affect the design and execution of testing activities.

This chapter focuses on the topic of Android mobile application testing. We discuss problems, questions, new challenges, and perspectives in the testing of these applications, including new techniques and tools. We divide the chapter as follows.

In Section 2, we provide a preliminary description about mobile applications and their evolution. In Section 3 we present an analysis of mobile application testing challenges and perspectives. Section 4 provides an overview of the Android platform and Android applications and Section 5 describes problems and solutions for testing an Android application at several levels. In Section 6, we present strategies for Android testing, and in Section 7 we consider non-functional requirement testing of Android applications. Section 8 presents the features of some popular tools and frameworks for Android testing. Lastly, Section 9 provides conclusive remarks and future perspectives.

2. EVOLUTION OF MOBILE APPLICATIONS

We can largely attribute the success of mobile applications in recent years to the improvements of networks and hardware advancements in the field of handheld devices. These devices have evolved considerably. While the first mobile phones simply allowed calls, today's smartphones often have the capabilities of personal computers. Manufacturers now equip devices with increasingly powerful processors, larger and faster memory modules, large screens with advanced touch functionality, full QWERTY keyboards, and high-speed wireless connectivity such as Wi-Fi and cellular data networks. Moreover, devices now integrate a wide set of hardware sensors as well, such as a GPS receiver, accelerometer, magnetic compass, and radio receiver.

Alongside the hardware advances, software development approaches for mobile devices have also changed significantly. Due to device limitations, software platforms for early mobile devices did not offer applications to any software abstraction mechanisms such as operating systems for hiding the hardware details of the underlying device. Conversely, modern smartphones exploit platforms including not only an operating system, but also application frameworks, libraries, and several other support tools, such as hardware virtualization utilities. This additional support allows a more viable development approach. For a more complete picture of the current platforms available to mobile application developers, Doernhoefer presents a survey on operating systems adopted by modern mobile devices in [41] and Gavalas and Economou review popular mobile platform development options for mobile devices in [45].

Fling [43] proposes the representation of the mobile ecosystem reported in Fig. 1. This model describes the modern mobile world as a multi-layer architecture that abstracts from the lower layers of network and device details up to the various software layers platforms now provide.

As shown, software layers above the device itself include Platform, Operating System, Application Framework, Applications, and Services. The Platform layer includes a core programming language and related tools (such as debuggers, emulators, and IDEs) offering a set of development APIs that

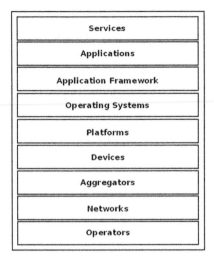

Fig. 1. The Mobile Ecosystem.

work similarly across multiple devices. Development platforms can be of three types: *licensed*, *proprietary*, and *open*. Device makers purchase licensed platforms (e.g., JavaME or Windows Mobile platforms) for non-exclusive distribution on devices. In contrast, device makers control the design and development of their own proprietary platforms (e.g., BlackBerry and iPhone), used exclusively on their devices. Finally, freely available open platforms (e.g., Android) allow users and developers more freedom to download, alter, and edit without limiting the user to a narrow set of devices.

The Operating system layer includes core services and toolkits that enable applications to interact and share data and services. Popular operating systems include Symbian, Windows Mobile, iPhoneOS (IOS), Palm OS, and Android OS.

The latter three levels of the architecture belong to Application Frameworks, Applications, and Services, respectively. Application frameworks offer developers an additional set of APIs and tools beyond the operating system itself, such as window and multimedia APIs. The Applications layer obviously includes applications deployed on the device. Finally, the Services layer includes applications intended to make a service available on the device, such as a Messaging, E-mail, or Web Browser application.

This layered software architecture enables new development approaches as hardware, operating systems, and applications (or theoretically, any layer) can be developed by separate parties. Moreover, developing mobile applications become increasingly similar to developing desktop applications, as mobile development requires fewer special skills. The boom of mobile application development in recent months coincides with these new opportunities. In particular, the availability of open platforms (like Android), the free availability of development tools, environments, and utilities, as well as the availability of application markets (such as the Android Market) for making applications available all allow for a wide community of "homebrew" developers [40]. These third-party developers contribute to the diffusion of small or even very small software projects that expand the functional capabilities of mobile devices and eventually contribute to their commercial success.

Bridging the gap between desktop computers and handheld devices represents the main challenge that research in mobile applications faces in the near future. According to Android guru Andy Rubin, "There should be nothing that users can access on their desktop that they can't access on their cell phone [65]."

Given this encouraging scenario for mobile applications, development processes must consider several relevant implications and challenges. According to Wasserman [76], these areas include: the definition of suitable approaches and principles for user interface design; finding effective solutions for achieving non-functional qualities in mobile applications (such as performance, reliability, usability, and security); defining development approaches and tools to support cross-platform development and testing; and finding effective and efficient solutions for mobile application testing.

Software testing remains one of the most critical and expensive activities in software lifecycle. However, the testing of mobile applications can be even more complex due to specific features and issues that characterize these applications. We analyze mobile application testing challenges and offer specific testing perspectives in the following section.

3. MOBILE APPLICATION TESTING CHALLENGES AND PERSPECTIVES

Many characteristics of handheld devices influence both the development and testing strategies of mobile applications.

Two main characteristics are the *heterogeneity* of the hardware configurations of mobile devices and the *variability* of their running conditions. Commercial mobile devices come equipped with different types and numbers of hardware sensors (such as GPS, magnetic compass, accelerometer, etc.), various types of screen displays (having different sizes, physical characteristics, and user interface capabilities), CPUs with different characteristics, and so forth. Moreover, devices support various mobile networks and communication technologies (like infrared, Bluetooth, etc.). These factors imply that mobile applications require cross-platform development and testing [54]. Due to the limitations of mobile devices, it is not practical to run a testing tool directly on the mobile device. Consequently, developers often carry out the testing of applications in two distinct steps. First, they test applications on a desktop or other development machine using emulators, then they must deploy and test on the target mobile device itself, since the emulators cannot assure complete compatibility with the target device [67].

The *scarceness of resources* of the hardware platform represents an additional challenge in modern mobile application development. Even with improvements in recent years, limitations include processing capability, memory restrictions, and power supply, as well as limited user interface features. These resource restrictions require specific testing activities designed to reveal failures in the application behavior due to resource availability. For example, tests must evaluate the application under conditions of high memory usage, low battery levels, and with several competing running processes, in order to expose possible failures due to extreme resource usage.

Similarly, possible *limitations of the wireless network* such as reduced bandwidth and connection instability also affect the user experience with mobile applications. As with the hardware resource limits mentioned above, specific testing activities such as mobility and usability testing should address expected network restrictions [67].

The exposure of mobile applications to *security* attacks also requires attention from testing activities. Due to the open nature of the mobile platform and marketplaces allowing users to easily download, install, and run applications, the sensitive data and powerful functions of the device are easy targets for applications running in the same environment. Security testing activities should reveal potential vulnerabilities of the mobile applications.

All of these factors produce new challenges for the quality assurance and testing process. Even though mobile applications require emphasis on certain activities, the testing of mobile applications builds on the general principles that rule any testing approach.

In particular, testing relies on the same well-known basic aspects [33], such as:

— *test models*, representing the relationships between elements of an abstract representation or an actual implementation of a software component;
— *testing levels*, specifying the different scopes of the tests to be run, i.e., the collections of components to be tested in each level;

- *test strategies*, defining heuristics or algorithms to create test cases from software representation models, implementation models, or test models;
- *testing processes*, defining the flow of testing activities, and other decisions regarding when testing should be started, who should perform testing, how much effort should be used, and similar issues.

Moreover, in order to cope with the inherent complexities and challenges of testing a mobile application, the testing process requires the consideration of multiple perspectives. For example, both non-functional and functional requirements require testing. The former perspective verifies the conformance of the application with specified (or even implicit) non-functional requirements like performance, usability, and security. The latter perspective tests the functional requirements of the application, such as its basic functionality. Clearly, testing a mobile application requires the consideration of both perspectives.

In the following sections, we analyze these basic aspects of testing with respect to the world of Android mobile applications. We begin with an overview of development on the Android platform.

4. OVERVIEW ON THE ANDROID PLATFORM AND ANDROID APPLICATIONS

The Android platform is currently one of the most popular open platforms for mobile devices. The Open Handset Alliance (actively supported by Google) develops the platform. The Alliance seeks to develop an open source mobile platform based on the Java programming language [27].

The Android platform contains of four layers, namely Applications, Application Framework, Libraries (including the Android Runtime), and a Linux Kernel. Figure 2 shows the software stack making up the Android platform.

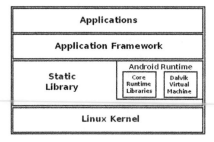

Fig. 2. The Android platform.

Applications occupy the upper layer of the stack, including core applications such as an email client, messaging program, calendar, maps, browser, contact manager, and others.

The Application Framework layer provides a set of reusable components for building new Android applications and includes services for applications (such as the Resource Manager, Notification Manager, and Activity Manager).

The Static Library layer contains a standard set of libraries and the Android Runtime. The Android application framework exposes the capabilities of these libraries to developers Capabilities include browser technology, database connectivity, advanced graphics and audio/video support, and SSL capabilities. The Android Runtime itself includes the core runtime libraries needed by the operating system and the Dalvik virtual machine (VM), an optimized, Android-specific Java virtual machine.

The Linux kernel provides a hardware abstraction layer and core services such as process, memory, and file-system management. The kernel layer includes hardware-specific drivers for the device as well.

Each Android application runs in its own process, with an individual user ID, with its own instance of the Dalvik virtual machine. This design provides separation between applications and provides a basic level of security.

An Android application can include four basic types of Java components: Activity, Service, Broadcast Receiver, and Content Provider. Offered by the Application framework, these components have different responsibilities and roles in Android applications.

The Activity components provide crucial functions for the application's user interface [5] and present a visual user interface for each specific task the user can undertake. An application usually includes one or more Activity classes that extend the base Activity class provided by the framework. The classes react to events generated by users and other system components. Additional framework classes (e.g., View, ViewGroup, Widget, Menu, and Dialog) provide the visual components associated with each activity on the screen.

In its lifecycle, an Activity instance passes through three main states, namely *running*, *paused*, and *stopped*. At runtime, just one activity instance at a time will be in the *running* state, having complete and exclusive control of the screen of the device. An Activity instance can make dynamic calls to other activity instances, and this capability causes the calling activity to pass to the *paused* state. A paused activity loses its exclusive control of the screen, but remains visible to the user. When a paused Activity becomes completely

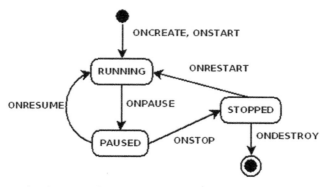

Fig. 3. The Android Activity lifecycle.

obscured by another Activity, the blocked Activity enters the *stopped* state. Figure 3 shows a simplified representation of the lifecycle of an Activity.

Applications may also include a second type of component—the Service. The Service component can perform long-running operations in the background (such as network transactions, playing music, performing file I/O, etc.) offering no user interface. Other components can activate services in two ways. Components launch *unbounded* services by an asynchronous service call, allowing the caller component and the service to proceed independently. *Bounded* services, in contrast, support inter-process communication (IPC) functions useful for synchronization and message exchange.

Components may launch Services can either for execution in a different process or thread, or in the same one as the caller. Running Services may stop themselves or be stopped by another component. A bounded service automatically stops when released by all binding applications.

The third type of component in Android applications is the Broadcast Receiver. This component listens and reacts to system-wide broadcast announcements that may originate either from the system (e.g., broadcasts announcing that the screen has turned off, that the battery is low, or that a picture was captured) or from other applications. Commonly, a broadcast receiver does a minimal amount of work, simply acting as a "gateway" to other components. For instance, it might initiate a Service to perform some work based on a received event. The Broadcast Receiver differs from the Activity component in that the Activity can listen for events only while running, but a Broadcast Receiver listens for events continuously.

Finally, the Content Provider component manages data for an application. Options for storage include the file system, an SQLite database, on the web, or in any other persistent storage location. Content providers also control the accessibility of an application's data by other applications.

One unique aspect of the Android system is support for runtime activation of components. In Android, any application can activate and reuse another application's component. For example, one application may use a photo-capturing Activity belonging to another application. However, because the system runs each application in a separate process and enforces file permissions that restrict access across applications, an application cannot directly activate a component from another application. Rather, the initiating application must deliver a message to the system that specifies an *intent* to start a particular component. The *Intent* class of the Android framework implements this capability. In particular, asynchronous intent messages activate three of the four component types described above—activities, services, and broadcast receivers. Intents also bind individual components to one another at runtime. As a consequence of this component activation mechanism, Android applications, unlike applications on most other systems, don't have a single entry point (i.e. no main() function or similar).

Partly because of having no concrete entry point, the Android platform enforces an interesting process management approach. When an application component starts and the application has no other running components, the Android system starts a new Linux process for the application with a single thread of execution. By default, all components of the same application run in the same process and thread (called the "main" thread). If a component starts and there already exists a process for its application then the system starts the new component within the existing process and in the same thread of execution. However, different components in an application can run in separate processes if desired, and additional threads for any process can be created by existing components.

On the Android system, every application must have an *AndroidManifest.xml* file that contains essential information about the application. This file declares the name of the Java package for the application, the components of the application (its activities, services, broadcast receivers, and content providers) with their names and capabilities (for example, which Intent messages they can handle). Additionally, the manifest file specifies which processes will host application components, the application's permissions to access protected parts of the API and to interact with other applications, as well as the permissions required in order to interact with the application's components. Finally, the manifest specifies the minimum version of the Android Runtime that the application requires and lists the libraries

Fig. 4. Components of the .apk Android Package.

required for linking against the application. The manifest file, then, specifies many details about how the levels of the Android system from Fig. 2 interact.

Developers compile and package Android applications down to a single Android application package file having the ".apk" extension. This file includes all of the application's code (.dex files), resources, assets, and manifest file. Figure 4 summarizes the main components of the.apk Android Package.

4.1 A Running Example

In this section, we present the characteristics of a simple Android application that we use as a running explanatory example for the remainder of the chapter.

The example application provides a simple Proximity Alert that will show a message on the device display and will activate the device vibrator when the device reaches a specified location of interest.

The application allows the user to configure the alarm by a "Configuration" user interface where the user can input the geographic coordinates of the desired location and activate the alert by pressing an "Activation" button.

Once the user activates an alert, the system listens in the background to GPS inputs, and when the device arrives at a distance of less than 30 m from the chosen location, the application will activate the vibration and show a notification message.

We implement this application with three main Android components. They include a "Configuration" Activity component that provides the functionality of alarm configuration and activation; a "Notification" Activity that shows the notification message; and a Broadcast Receiver component that listens for Intent messages with location information.

The "Configuration" Activity is the only component directly launched by the user. The Configuration GUI includes two editable text fields for entering the latitude and longitude values of the location of interest, and two buttons—the "Activation" button and the "Set Current Coordinates" button—and four text fields showing labels or messages. The Android classes used for implementing these widgets include EditText, Button, and TextView, which all extend the AndroidView class.

When the user clicks the "Set Current Coordinates" button, the application reads the current coordinates (the last ones obtained from the GPS receiver) and uses their values to set the location input fields of longitude and latitude. Users can then modify these values manually. When the user clicks on the Activation button, the Configuration Activity instantiates the Broadcast Receiver and then stops. A screenshot of this activity is shown in Fig. 5.

The Broadcast Receiver acts as a proximity sensor, in the sense that it listens to the Intent messages generated by the Android system from the GPS receiver of the device communicates a position within 30 m of the location of interest. When the receiver communicates a position within 30 m of the location of interest, the Broadcast Receiver component activates the Vibration and the Notification Activity components in parallel.

Fig. 5. A screenshot of the Configuration Activity.

Fig. 6. A screenshot of the Notification Activity.

Fig. 7. The Class diagram showing the components of the Proximity Alert Application.

The notification activity shows a user interface containing just a single text field with the string "You reached the goal!" and a button to dismiss the notification activity itself. Figure 6 shows a screenshot of this user interface.

The class diagram in Fig. 7 describes the components of the application.

The indirect dependency relationship between the Configuration class and the ProximityIntentReceiver class depends on an instance object. The Configuration activity actually instantiates the Intent object specifying the proximity to the chosen location, though the operating system itself will asynchronously send this intent message to the Broadcast Receiver as the device approaches the location. The ProximityIntentReceiver activates the Notification Activity directly.

```
<uses-sdk android:minSdkVersion="3"/>
<uses-permission android:name="android.permission.VIBRATE"></uses-
permission>
<uses-permission
android:name="android.permission.ACCESS_COARSE_LOCATION" />
<uses-permission
android:name="android.permission.ACCESS_FINE_LOCATION" />
<uses-permission
android:name="android.permission.ACCESS_LOCATION" />
<uses-permission android:name="android.permission.ACCESS_GPS" />
```

Fig. 8. The section of the Android manifest file of the Proximity Alert Application specifying the permission requests of the application.

Besides other information, the manifest file for the application specifies by the *android:minSdkVersion* attribute of the *uses-sdk* tag that the application requires Android version 1.5 (named "Cupcake"), supporting the version 3 of Android APIs. (In practice, no devices currently on the market depend on versions prior to Cupcake.) The manifest file also specifies by the *uses-permission* tag that the application needs access to the vibration feature and location services of the device. Figure 8 contains an excerpt of the manifest file.

5. ANDROID APPLICATIONS TESTING

Constrained by the specifics of the Android platform discussed above, Android application testing aims to execute the application using combinations of input and states to reveal failures. By failure, we mean the manifested inability of a system or component to perform a required function within specified performance requirements [48]. Sources of failures include faults in the application implementation, the running environment, and the interface between the application and its environment.

This classification of faults aligns with the findings of a recent study of the Android operating system that characterized about 630 bugs found in real Android applications [55]. This study revealed that most bugs were located in application implementations while the remaining bugs were concentrated in the lower layers of the Android platform (e.g., kernel and device drivers).

Hu et al. [47] present another interesting classification of Android bugs. They derive a classification scheme from an analysis of bugs in 10 free Android applications from the Android Market. The resulting scheme distinguishes between seven types of bugs, such as bugs from activities, events, dynamic type errors, I/O and concurrency errors, unhandled exceptions, and others. The analysis shows that, while bugs in application logic still occur, the activity and event-based nature of Android applications contributes to the remaining bug classes. This finding further motivates the need for Android-specific testing procedures, such as those we detail throughout this chapter.

Testing processes usually exploit a combination of testing strategies and consider several testing levels in order to reveal the different types of observable failures in software applications. Testing strategies define the approaches for designing test cases. Possible strategies include black-box, white-box, and gray-box techniques. Black-box strategies design test cases based on the specified functionality of the item to be tested. White-box strategies consider the underlying source code to develop test cases, and gray-box strategies design test cases using both functionality- and implementation-based approaches [33].

Testing levels define the different scopes and points of entry for tests. Unit testing verifies each individual source code component of the application, while integration testing considers combined parts of an application to verify their combined functionality. Finally, system testing aims to discover defects by testing the system as a whole rather than testing individual or smaller combinations of components. In a typical bottom-up approach, the testing process begins with unit testing, followed by integration testing and ultimately system testing.

In this section, we analyze specific problems in the testing of Android applications presenting testing guidelines, techniques, and technologies for each of the unit, integration, and system levels.

5.1 Unit Testing

In the context of Android testing, three preliminary questions about unit testing need to be addressed. First, what is the scope of Android unit testing? Second, what are the most effective strategies for Android unit test case design? Finally, how can we automate Android unit testing?

For the problem of scope, Android applications contain different types of units, including both typical Android components (like activities, services, broadcast receivers, and content providers) and components implemented by standard Java classes. The simplest solution consists of carrying out unit testing on both framework-based and ordinary Java classes not belonging to the Android framework.

Regarding testing strategies, developers can still apply all three types of techniques to achieve testing goals on Android applications [33]. However, the complete testing process must consider the event-driven behavior of the Android framework components and lifecycle-related peculiarities, possible requiring some *ad hoc* or exploratory test cases as well.

By testing automation, we mean the set of techniques and technologies used for automating and supporting the typical tasks of a testing process [72]. Testers apply automation to many activities, such as defining and running test cases and test suites and evaluating the testing results. To this aim, the Android development platform includes the Android testing APIs, which offer a rich set of Android-specific testing classes similar to the ones offered by the JUnit testing framework for Java applications [49]. As an example, these APIs provide classes for defining and executing test cases for each type of Android application components, as well as assertions for checking testing results, and even support for stubs, fakes, and mock classes for testing single classes in isolation from the external context. Moreover, the testing framework includes classes for the automatic instrumentation of application code and the monitoring of the interactions between the system and the application, as well as classes for emulating the Android lifecycle events. Finally, the framework includes classes for initializing application context (e.g., application-specific resources, sensors and system services) and isolating a tested component from the real execution context. This extensive support for writing tests within code directly supports unit test automation on the Android platform.

In the following subsections, we analyze unit testing techniques for Activity, Service, Broadcast Receiver, and Content Provider components.

5.1.1 Activity Testing

The unit testing of Activity components provides an Activity component with a given input in isolation from the remaining parts of the application and checks the resulting behavior of the component. An Activity triggers behavior in response to three different types of events: events related to its lifecycle, user events, and systems events (such as sensor, time, and other events external to the Activity). Consequently, Activity testing should consider the following specific tasks:

1. Test the Activity's response to its lifecycle events.
2. Test the Activity's response to user events.
3. Test the Activity's response to system events.

At the lowest level (i.e., white-box testing), testers can use traditional assertion-based techniques to check the behavior of the Activity. For example, assertions can check properties of the UI widgets at any point during a test case, because the Activity already manages user interface details.

Hu et al. [47] propose a technique for testing the Activity's response to lifecycle events. This technique exploits the state machine-based representation of the activity lifecycle reported in Fig. 9. The state machine defines

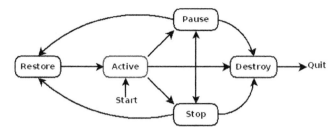

Fig. 9. Simplified state machine of an Android activity.

the correct behavior of the activity in terms of possible paths on the state machine. Hence, violations of the state machine represent potential bugs in the application. The technique uses system log files to detect violations.

More precisely, the testing activity will have to check that:

1. the sequence of method calls changing the Activity state complies with the state machine transitions of an Android activity;
2. the overall state of the application is preserved when the Activity resumes after being interrupted.

For example, when the device orientation changes (from vertical to horizontal and vice versa), the user interface layout must change too (from portrait to landscape, or vice versa). In this scenario, the running activity moves to a paused state, and must redraw its user interface according to the new device orientation when allowed to resume. Similarly, a state transition from Active to Pause occurs each time the mobile device receives a phone call or message.

Testers can perform this type of Activity testing using the Android testing framework instrumentation capabilities, which provide methods for emulating Activity lifecycle events. For example, the method *callActivityOnPause* brings an Activity into a paused state. Similarly, the analogous *callActivityOnStart*, *callActivityOnStop* can emulate transitions for Activity testing. Moreover, the *IsolatedContext* class allows a tester to set up a specific execution environment of application-specific resources and classes.

In addition, testers must test an Activity against specific user events, i.e., events fired on the User Interface shown by the Activity on the device display. As in previous scenarios, this testing aims to verify the correctness of Activity behaviors. Tests should check for *event errors* that occur when the application reacts to an event by performing a wrong action [55].

One possible strategy to define test cases for user events considers all the "clickable" components of the user interface managed by the Activity (i.e., registered widgets associated with specific user events). Complete user testing will require consideration of event sequences of various lengths, which must be defined. As an example, testers must pay particular attention to event sequences representing specific user tasks. An example of such a sequence may be composed of the following events: (1) filling in a text field, (2) selecting a check button, and (3) pressing a "submit" interface button.

As to the automation of user event testing, the Android testing framework [23] offers the *ActivityInstrumentationTestCase2* class for functional testing of Activity classes and methods. This class can emulate user events and solicit the Activity in isolation from other components. More precisely, the *setActivityInitialTouchMode* method of the *ActivityInstrumentationTestCase2* class can disable the sensitivity of an Activity to real user events, while the *sendKeys* method simulates click events on the device keys (such as menu, home, back, direction keys, and so on). Analogously, the *TouchUtils* class provides methods to simulate user interaction with the application by means of a touch screen.

Lastly, we report some considerations for testing an Activity's response to system events. The operating system generates these events in response to solicitations from external sources, such as device sensors, timers, batteries, USB drivers, network details, and other sources within the system. Activities can raise system events asynchronously and provide handlers for managing these events.

The testing of an Activity with respect to system events can be much more difficult than testing with respect to user events, due to the wide set of sensors and system services available on modern mobile devices. Consequently, this type of testing requires suitable strategies for reducing the number of tests necessary. To this aim, the tester should utilize the application's manifest file. Originally designed for security purposes, the Android manifest specifies the system services employed by the application and helps the tester reduce the number of system that tests should fire on the subject Activity objects.

In order to automate the test case implementation and running tasks, test cases must programmatically emulate system events. The *android.hardware* package contains classes which support this emulation. For example, in order to emulate sensor events, tests can use classes like *Sensor, SensorEvent, SensorEventListener,* and *SensorManager*. A tester has to modify or extend these classes in order to test the Activity with respect to fictitious events generated by sensors.

Sensor Simulator [71] presents an example Java program able to emulate physical events such as movements and accelerations of the device.

5.1.2 Service Testing

Application components of many types (e.g., Activity, Broadcast Receiver, Service, or Intent messages) can launch Service components. The Service object can be classified as *unbounded* (similar to a function with a single entry point that returns no data to the caller) and *bounded* (that can additionally interact with their caller by inter-process communication mechanisms). A Service runs in the background of an application and supports tasks such as long-running computations, music playing, and downloading.

Service components have a simple lifecycle. The service deploys after its creation via a *StartService* call (for unbounded services) or a *BindService* call (for bounded services). Unbounded services may stop themselves or their creating components may stop them. At this point, the operating system shuts down and destroys the Service. The OS destroys bounded services only when all its clients release it with *onUnbind* calls [21]. Figure 10 shows the lifecycle of both types of Service components.

Analogous to the Activity testing procedures discussed above, Service testing focuses on the correct management of lifecycle events. Additionally, Service testing should assert the correctness of any business logic implemented by a Service. Testers can apply black-box techniques to test business logic, including applying input partitioning to test the Service with respect to both valid and invalid input values.

As to the automation of Service testing, the Service lifecycle events can be emulated by methods of the *Context* class [12], such as *startService, bindService,*

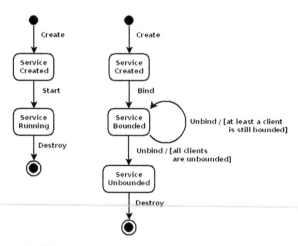

Fig. 10. The Service lifecycle.

and *stopService*, that allow test cases to start and stop bounded and unbounded services directly. Test cases can isolate Service components with the *Isolated-Context* class, which substitutes the actual context of execution for a service and other elements of the application. Methods *setContext* and *setApplication* of the *ServiceTestCase* class allow the tester to link tested services in an isolated context.

5.1.3 Content Provider Testing

Content Provider components virtualize access to shared data resources, such as local or remote databases. These components offer methods for data access and modification such as query, insert, delete, and update methods [11]. Any Content Provider has a unique URI for identification. Other components access the Provider by its URI.

To unit test Content Provider components, test cases only need to interact with the Provider's methods. In order to execute these test cases, virtual data sources should substitute real sources.

The Android testing framework, as in other cases of unit testing, provides classes to support unit testing of Content Provider components, such as *ProviderTestCase2*. The previously cited *IsolatedContext* class and the *Mock-ContentResolver* class allow test cases to redirect a data request of the Content Provider to a mock provider implemented for testing purposes.

5.1.4 Broadcast Receiver Testing

Broadcast Receiver components listen for Intent messages and can execute basic computations in response. These components usually call other activities or services for Intent handling. At the unit testing level, the testing of a Broadcast Receiver component requires the verification of the business logic provided by this component upon receipt of each registered type of Intent message the component may receive.

To this aim, test cases must simulate Intent messages and assert the correctness of the business logic implemented by the Broadcast Receiver. Test cases may artificially send an Intent message to a Broadcast Receiver using the method *sendBroadcast* provided by the *Context* class of the Android testing framework.

5.2 Integration Testing

Integration testing builds on unit testing to test sets of related application components in order to assess how they work together and to identify failures due to their coupling.

Adequate integration strategies and criteria define the set of components to be integrated and tested as a whole. In traditional software systems, testers usually consider three main integration strategies: bottom-up, top-down, and sandwich [72].

In bottom-up integration testing, testing considers the lowest level of components of the application first and then uses these components to facilitate the testing of higher-level components. The process repeats, moving upward to eventually test the components at the top of the application's hierarchy. Conversely, top-down approaches proceed by first testing the higher-level components, and working down to test lower-level components. The sandwich strategy is a hybrid strategy that mixes top-down and bottom-up approaches, presumably in a way that follows from the application's design or specific testing needs.

In Android client-server applications, a possible integration approach may be to start by integrating the client-side components of the application and then work hierarchically to the server-side components. This approach increases the likelihood of isolating a bug to the client or server-side of the application.

In client-only application, the tester can focus just on integrating the inner components of the client. Considering the client side in isolation, if design documentation shows static and dynamic dependency relationships between software components, this documentation can guide integration testing. In fact, the Android Manifest file included in the APK details the components of the application and the main static relationships between components. However, because an Android application is actually an object-oriented program implemented in Java, testers must also consider dynamic dependency relationships among classes. Additionally, dependency relationships due to specific late binding mechanisms of Android (such as the ability of Intent messages to implicitly interact with components) further complicate component dependencies—and subsequently, integration testing.

In integration testing, testers must distinguish between lower-level and higher-level components of the application. Lower-level components will be terminal components in the application's hierarchy that do not depend on any other components within the application. Conversely, higher-level components are much more likely to depend on other components. Service and Content Provider components make great candidates as terminal components, since Content Providers usually depend on the external data sources they access or on services they call. Unbounded services may depend only on other unbounded services, while bounded services may interact with

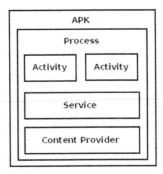

Fig. 11. Android Software Components in the APK.

activities or other services, too. Because activities directly handle user and system events, they make better candidates for higher-level components.

Figure 11 shows a possible organization of components for an Android application as deduced from the application's APK.

Once a tester discovers the application's hierarchy of components and its dependency relationships, integration testing can proceed in any reasonable fashion. In bottom–up approaches, the integration will start from the lower-level components (likely Content Provider or Service components). In top–down approaches, integration starts from entry-point components, such as Activity components (in particular, the manifest file declares the only directly launchable Activity as *MAIN*), Service components running in background, or Broadcast Receiver components not started by other components of the application. The top-down integration can also start from AppWidgets too, which are extensions of Broadcast Receiver classes, and provide further entry points to Android applications. The testing approach must consider any such extensions of the base components.

The sandwich integration testing strategy consists of a mix of the two previous strategies. While not likely as clean as the previous two strategies, applications with strongly connected components sometimes require this hybrid approach to overcome a lack of a clear hierarchy between components.

5.3 System Testing

IEEE defines system testing as "the testing conducted on a complete, integrated system to evaluate the system's compliance with its specified requirements" [48]. System testing will check for compliance with both functional and non–functional requirements and will aim to discover failures most likely due to the underlying device and the real execution environment of the mobile application.

The heterogeneity of mobile device hardware platforms, among other factors, makes system testing of mobile applications a non-trivial process [77]. In general, testers must run applications on an array of different devices with various resource limitations. Mobile system testing generally proceeds in consecutive stages in order to control for heterogeneous features of devices. Testers first test the system using virtual machines and sophisticated simulation environments, and then proceed to replace virtual devices and environments with real ones iteratively.

In this perspective, mobile system testing shares many problems and solutions with embedded software testing. Testing embedded systems software usually proceeds in three main stages called *simulation*, *prototyping*, and *pre-production* stages [36]. In the simulation stage, test cases run in a completely simulated environment. In the prototyping stage, the actual components gradually replace the simulated components while the embedded software still executes on an emulated processor. Finally, in the pre-production stage the embedded software executes on a real processor and in a real environment. Different levels of effort, cost, and needed time characterize each stage. For example, the simulation stage supports easier and inexpensive test design and execution, while tests in pre-production stage can involve the use of some limited and costly resources.

In the following subsections, we discuss how Android application testing can adopt this staged approach to system testing.

5.3.1 Simulation Stage of System Testing

The simulation stage of Android system testing consists of running an entire application with programmatically described test cases in an *isolated* environment on an *emulated* device.

The real execution environment of a mobile application normally includes several types of interaction and real input, such as user events, network signals, GPS signals, etc. At the simulation stage, the isolated environment of the running application will include fictitious software components (e.g., mocks and stubs) that provide emulated input to the application. At the same time, the application will run on a software platform that emulates the real device rather than a real physical device.

In particular, the Device Emulator software included in the Android Development Toolkit provides Android device emulation [8]. The Emulator allows the development and testing of Android applications without a physical device. The software consists of an executable program runnable on a host computer that can emulate any virtual device, including user-configured

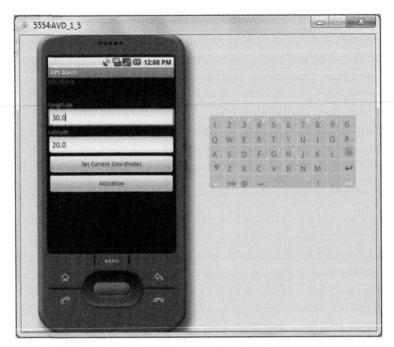

Fig. 12. The Android Emulator showing an emulated device.

Android operating system and hardware characteristics. The GUI of this emulator shows a faithful reproduction of the screen and keys of the emulated device. The emulator also allows interaction with the device by simulating the touch controls by means of mouse interactions and keyboard key presses. The simulation phase does not require interaction with the emulated device, but the prototyping step will use these features. Figure 12 shows the GUI of the Emulator while displaying an emulated device.

An Android Virtual Device (or "AVD") file defines a virtual device for emulation by the Android Emulator [9]. The AVD file describes the hardware and software options of the emulated device. Interestingly, Google released the Android Emulator on November 2007, before the release of any real Android-powered device, so that the first Android applications were completely developed and tested in the emulated environment [40].

5.3.2 Prototyping Stage of System Testing

In the prototyping stage, the testing activities aim to exercise the application in a modified execution context that is not yet the real execution environment, but progressively approaches that environment. The application remains on the device emulator, but the environment progressively removes isolation.

Consequently, this stage will involve several consecutive iterations of system testing. The tester progressively replaces isolated features with real environment features, sometimes implemented by means of mock testing classes (e.g., sensor mocks, GPS mocks, etc.). For example, the tester will remove classes emulating interaction with the user interface in favor of executing test cases directly on the Emulator GUI. Analogously, the *Context* class, which implements the real context of the device (in this case, the emulated device) will replace the *IsolatedContext* class of the Android testing framework, which isolated components with mock contexts.

The Dalvik Debug Monitor Server (DDMS) utility [26] included in the Android Development Toolkit allows the spoofing of location position changes, incoming calls and SMS messages. The DDMS actually supports interactive debugging by providing logging and debugging features, as well as features for monitoring phone status, active processes and threads, memory allocation, and file systems. The DDMS utility builds on the Android Debug Bridge (adb) utility [7] from the Android Development Toolkit. The command-line adb utility provides features for handling communication between virtual or real device instances and the emulator.

Figure 13 shows a screenshot of the DDMS utility in the Eclipse IDE. The first panel on the left (labeled "Devices") shows details about instantiated virtual devices and the currently active processes. The "Emulator Control" panel shows the interface that allows manual injection of GPS coordinates. The bottom "LogCat" panel shows the log of all recent events fired by the system and the running applications. Finally, the right panel contains three tabs reporting the list of application Threads, the Allocation Tracker, and an Explorer of files present on the device, respectively.

In the last steps of the prototyping testing stage, real target devices will substitute the Android Virtual Devices. The real devices physically connect via USB to the machine hosting the adb. In this way, tests run in a real environment including the real phone or the touch pad. However, tests requiring a real interaction with specific components of the device, like the GPS, accelerometer, magnetic compass, or other sensors, will be executable only on the actual device during Pre-production system testing.

5.3.3 Pre-Production Stage of System Testing

In the pre-production stage, system testing occurs in a real execution environment using a real device instead of a device emulator.

Fig. 13. A screenshot showing the execution of the DDMS utility in the Eclipse IDE.

This stage has the highest testing difficulty, as testers must produce testing scenarios from the real world. Tests may depend on several scenarios and require reproducible input in order to check application functions or other emerging qualities not available at lower levels of testing, e.g., performance with respect to various profiles of available resources.

The lack of debugging or monitoring facilities in the real execution environment, aside from any probes or logging facilities in the application itself, further increases the difficulty of testing in this stage. Even if such facilities exist in pre-production system testing, their presence may affect test results and will not likely persist in the final application.

5.4 Testing the Running Example Application

In this section, we present possible solutions for testing the functionality offered by the "Proximity Alert" example application described earlier in this chapter. We focus on testing this application at the unit, integration, and system levels.

5.4.1 Unit Testing

A test plan preliminarily defines the testing activities for execution on a component or application. At the unit testing level, we provide the following test plan for each component in our application:

1. Test the component behavior with respect to lifecycle events.

2. Test the business logic of the component with respect to invalid input.

3. Test of the business logic of the component with respect to valid input.

The Proximity Alert example application includes a Broadcast Receiver component named ProximityIntentReceiver and two Activity components named Configuration and Notification, respectively. The Notification Activity has just a single button that, once pressed, ends the activity. Because this component's unit testing is simple, we focus the following discussion on just the testing of the Configuration Activity and ProximityIntentReceiver broadcast receiver.

In order to verify the Configuration Activity behavior in response to events related to its lifecycle, we consider sequences of system events that cause the transition to the Pause state or resume the Activity from the Pause state. These transitions may be due to notifications from other applications, display orientation changes, or other potential causes.

As an example, the Configuration activity has to maintain the user interface state (given by the values of the latitude and longitude fields inserted by the user) when passing from the Pause state to Resume. We can replicate this transition with the sequence of events *onPause* and *onResume*. The example in Fig. 14 shows a possible JUnit test case that tests the Activity's response to this sequence of events. The test case initially sets the latitude editText field with a value of 45, then the activity is paused and resumed. Upon resume, an assertion checks that the latitude value still equals the expected value of 45.

```
public void testcaseLifecycle(){
    setActivityInitialTouchMode(false);
    myActivity=getActivity();
    Instrumentation mInstr = this.getInstrumentation();
    latitude.setText("45");
    mInstr.callActivityOnPause(myActivity);
    mInstr.callActivityOnResume(myActivity);
    assertEquals(latitude.getText().toString(), "45");
}
```

Fig. 14. A JUnit test case checking the behavior of the configuration activity in response to pausing and resuming the activity.

```
public void testcaseInputValidation(){
    setActivityInitialTouchMode(false);
    latitude.setText("100");
    activation.performClick();
    assertEquals(message.getText().toString(),"Incorrect
    values");
                                    }
```

Fig. 15. A JUnit test case related to the validation of the latitude text field.

The second type of unit test will check the component behavior when it receives invalid user input.

As an example, the Configuration Activity should check the validity of longitude and latitude input values before using them. Valid latitude values must be of numeric type and belong to the range from -180 to 180, while valid latitude values must belong to the range from -90 to 90. A possible test case for invalid input, then, will set either latitude or longitude to an invalid value, and try to activate the alarm, asserting whether the activity displayed a warning message. Figure 15 shows such a test case.

In addition to the testing latitude and longitude values (including tests with valid values), further tests must also check the Activity's business logic with respect to GPS input, since the Configuration Activity also receives GPS input.

A first test case will check the behavior of the application when the GPS receiver is disabled (and the user presses the setCurrentCoordinates button).

A second test case will assert correct reporting of the current coordinates into the latitude and longitude fields of the GUI after the clicking of the setCurrentCoordinates button. This test case will check the resulting behavior of the activity when fictitious GPS coordinates are injected in the input fields. This unit test will require the implementation of mock objects and additional components. Figure 16 shows the class diagram with the classes involved in the definition of such a test case. The diagram includes the following classes:

– Configuration Activity—the class under test.
– TestClass—extends the ActivityInstrumentationTestCase2 class provided by the Android testing framework and includes the basic test methods needed by JUnit tests (such as *setup* and *tearDown*) and the test method *testcaseGPS*. The TestClass includes a LocationManager object (*mylocman*), too, linked to the MockProvider class.
– MockProvider—instantiates an artificial provider of geographic coordinates values. Coordinate values can be set with the *setPosition* method.

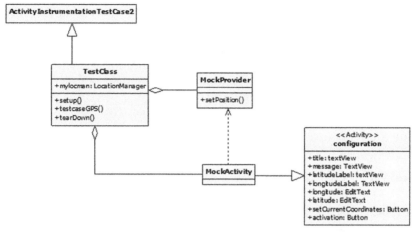

Fig. 16. Class diagram showing the classes needed for testing the Configuration Activity with a fictitious GPS Provider.

– MockActivity—extends the Configuration Activity under test to override its onCreate methods in order to force the Activity to use the MockProvider instead of the real GPSProvider. The TestClass will directly interact with this class instead of with the Configuration Activity.

This set of classes supports the creation of test cases that test the Configuration Activity by means of artificially generated geographic coordinates from the TestClass itself, via the MockProvider. A test case can set arbitrary coordinates with the MockProvider class and use assertions to check their handling by the Configuration Activity.

Lastly, we describe the testing of business logic in the Proximity IntentReceiver Broadcast Receiver component.

This component has very simple behavior: the Configuration Activity instantiates the Broadcast Receiver, which registers to the PROXIMITY_ ALERT Intent message. When the component receives this message, it should activate the device vibration and call the Notification Activity. To test this behavior, one possible test case could generate the PROXIMITY_ALERT Intent message and then assert if the Broadcast Receiver correctly calls the Notification Activity.

5.4.2 Integration Testing

To carry out integration testing, we must preliminarily define an integration strategy. In this simple example, both the bottom-up and the top-down strategies apply. Using the bottom–up approach, we start testing from the lower-level components of the application and test the interactions between these components and the higher-level components.

In this case, starting from the terminal Notification Activity component, we first test the interaction between this component and the Broadcast Receiver, which calls this Activity. Because this behavior requires no other processing or data exchange, the test case must only assert that the Broadcast Receiver correctly starts the Notification Activity. We can use test cases already designed for the unit testing of these components to implement this integration test.

As integration testing proceeds upward, we must integrate the Configuration Activity with the other components to be tested as a whole. New integration test cases should assert the correctness of the Configuration Activity's instantiation of the proximity alert and setting the chosen location coordinates. These test cases can reuse much of the same structure developed for unit testing, such as the MockProvider class used to inject artificial geographic coordinates.

Should we choose a top–down integration strategy, the higher-level components integrate with lower-level components. In the case of our application, we would first integrate the Configuration Activity and the Broadcast Receiver before integrating the Notification activity.

5.4.3 System Testing

At the *simulation* stage of system testing, we execute tests of the whole application on the Android Emulator with a completely emulated environment. This environment does not include any real device sensors, but just their emulated versions, as provided by mock objects.

At the *prototyping* stage of system testing, tests will be still executed on the emulator, but more realistic input will be manually provided to the application by means of the GUI of the emulator software and the DDMS utility. This utility allows us, for example, to spoof GPS data in order to test our application with real GPS input. Further prototype-stage system testing will require the deployment of the application onto a real device while tests still exercise the application through the emulator. For our application, this step allows us to evaluate the device vibration functionality.

Finally, in the *pre-production* stage of system testing, the application runs on the real device and in a real environment. At this point, we can evaluate mobile test execution, with real data provided by the actual GPS receiver of the device.

6. TESTING STRATEGIES

Testing strategies define the approaches for designing test cases. We can categorize testing strategies broadly as responsibility-based (also known as black box), implementation-based (or white box), or hybrid (also known as gray box). A strategy can target various levels of testing, such as unit, integration or system testing.

Recently, researchers have presented some strategies for testing Android applications in the literature. Some *event-based techniques* test the GUI of Android applications [47]. Other approaches, called *GUI-based techniques*, use the GUI to define system tests [4, 74]. Moreover, researchers have applied techniques for random testing and *adaptive random testing* to Android system testing as well [53].

In the following section, we analyze these Android testing strategies in detail.

6.1 Event-Based Testing Strategies

Android applications actually fall into the much larger category of Event-Driven Software (EDS) applications—applications driven by several types of events. Like EDSs, they take user-generated and/or system-generated events as input, change their state, and (optionally) output an event sequence. Writing EDS typically centers on the implementation of a collection of event handlers designed to respond to individual events.

Several factors make the testing of EDS complex. A first question considers the definition of test oracles in order to determine whether a test case applied to the application under test has run as expected [58, 60]. Additionally, defining a representation of an EDS's test cases as sequences of events [78] and finding effective techniques for automatically generating and automatically EDS test cases [60] both present challenges. The typical approach used for the automated test case generation in EDS involves the creation, automatic or not, of an abstract model (e.g., state machine model [69, 78] or event-flow model [35, 59]) of the application under test (AUT) and using the model to generate test cases [80]. Given this approach, researchers currently consider the definition of suitable reverse engineering techniques for obtaining these models as part of an EDS testing process [1, 57, 61].

Researchers propose techniques in the literature for testing several types of traditional EDS systems, such as GUI-driven software [32, 80], Rich Internet Applications [2, 3, 56, 61], and embedded software [36], among others.

We can adapt these techniques from the testing of traditional of traditional EDS systems to Android mobile applications by considering the context of Android applications.

Hu et al. propose an interesting strategy for Android-specific event-based testing [47]. As mentioned previously in this chapter, the authors develop a classification of Android-specific bugs. The classification drives their event-based testing approach. Recall that the bug classification includes several types of defects that can be found in Android applications, such as Activity, Event, dynamic type, API, I/O, and concurrency errors, as well as unhandled exceptions.

The testing technique focuses on Activity, Event, and dynamic type errors. The tester generates test cases for each Activity of the application under test, exploiting the Activity testing classes from the Android testing framework that work in conjunction with JUnit. The technique uses many features from the Activity testing classes, including: *Initial condition testing* (that tests the activity's proper creation), *GUI Testing* (that checks whether the activity performs correctly according to the GUI's specifications), and *State Management Testing* (that tests whether the application can properly enter and exit states).

The test case generation centers on Activity components, as these components provide the main entry points and control-flow drivers in Android applications. The *Monkey* event generator generates random or deterministic sequences of events automatically, and these sequences define test cases [24]. *Monkey* also supports interaction with the mobile device.

Testers execute the test cases on the application as a whole through the Dalvik Virtual Machine, which logs the details of each test case into a tracing log file. Finally, the technique automatically analyzes the obtained log files in order to detect potential bugs by looking for known patterns. Each class of errors (activity, event, or dynamic type) has an associated pattern.

6.2 GUI-Based Testing Strategies

Takala et al. present a model-based testing approach for testing the GUI of an Android application [74]. Model-based testing (MBT) approaches first define a formal model describing the application at a level of detail necessary for automatic test case generation. Test case generation algorithms process the model in systematic ways to produce test cases.

The MBT technique proposed by the authors describes the GUI of an Android application by state machines, a very common model for representing GUIs. The state machine is made up of states, transitions between states, actions attached to transitions, and labels attached to states. In order to cope with the complexity of state machines representing real-size applications, this technique divides the GUI models into smaller model components. Each model component corresponds to an individual view of the GUI and subdivides into two levels as specified by two separate state machines: an action machine and a refinement machine. The action machine describes high-level functionalities using *action words* and *state verifications*. An action word, modeled as the action associated with a transition, describes a small use case of the application such as saving a file. A state verification, modeled as a label in a state, describes the state of the application in order to verify that the actual application state corresponds to the state of the model. The refinement machine describes the action words and state verifications using keywords. The tester must manually generate both models.

The test case generation relies on models and keywords. Since some keywords depict typical user events while other keywords verify the state of the application, keywords can be used to define and execute test cases with a test automation tool.

A set of testing tools called the TEMA tools supports this MBT technique. The tools aid different phases of the technique's process, such as test modeling, test design, test case generation, and test debugging. TEMA contains a tool for model design as well as other model-related utilities. Test design tools include a Web GUI for designing test objectives. The test generation tools provide a number of algorithms that use models and test objectives to produce actual test cases. Test debugging tools in TEMA help to interpret unsuccessful test runs. Finally, the only Android platform-dependent tools in the toolset support automatic execution of test cases.

The first three authors of this chapter propose an alternative approach for testing an Android application automatically by its GUI [4]. This automatic testing technique uses a GUI crawler that simulates real user events on the user interface of the application to infer a GUI model automatically. The GUI model supports the derivation of test cases that can be automatically executed for different aims, such as crash testing and regression testing.

The crawler produces a GUI model that is actually a tree structure called a GUI Tree. The nodes of the tree represent individual user interfaces in the Android application, while edges describe event-based transitions between interfaces. The crawler obtains the tree by dynamic analysis, reconstructing a model of the GUI by firing events on the application user interface. The model describes each user interface in terms of its component widgets, widget properties, and event handlers, while describing each transition between

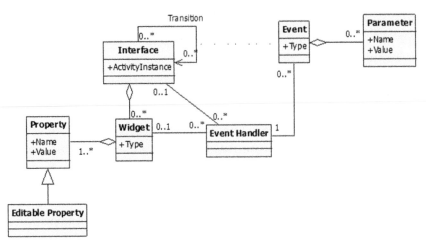

Fig. 17. Conceptual model of the GUI Tree.

consecutive user interfaces by the event that causing the transition. Figure 17 shows this model of an Android GUI as a class diagram.

The termination criterion of the GUI exploration activity represents a critical aspect of any GUI crawling algorithm. A frequently used criterion evaluates the equivalence of user interfaces. GUI exploration stops when the current interface appears equivalent to an already-visited interface. The crawling algorithm proposed in [4] assumes two interfaces to be equivalent if they have the same ActivityInstance attribute (see the model in Fig. 17) and the same set of Widgets with identical Properties and Event Handlers.

Another relevant aspect of any GUI crawling algorithm consists of finding an approach for defining the properties of widgets (such as the value of an input field) that must be set before firing an event on the GUI. As an example, the crawler presented in [4] assigns random values to properties.

Figure 18 shows a GUI Tree of a real example Android application.

The GUI tree generated by the crawler provides a starting point for automatic crash testing. According to Memon and Xie crash testing aims to reveal application faults due to run-time uncaught exceptions [59]. To execute crash testing, the technique presented in [4] uses test cases made up of sequences of events that correspond to GUI tree paths (from the root node to the leaves of the tree). Code instrumentation and monitoring support automatic detection of crashes during test case execution.

The same test cases generated for crash testing may also be used for regression testing. Testers perform regression testing after making changes to a given application under test. Regression testing consists of rerunning

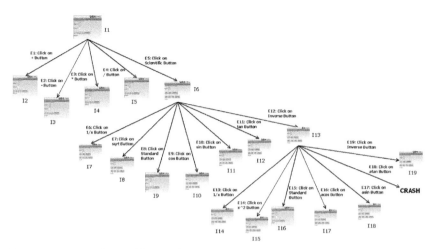

Fig. 18. GUI Tree obtained by crawling an example Android Application.

previously run tests and checking for changes in program behavior and the emergence of new faults. To detect differences, testers can compare the sequences of user interfaces encountered in both the test runs. The interface comparison can be made using test oracles having different degrees of detail or granularity [60]. As an example, the *Monkey Runner* tool [18] executes regression testing of Android applications but checks results just by comparing the output screenshots to a set of screenshots accepted as correct. The authors of [4] propose checking whether all intermediate and final interfaces obtained during test case execution coincide with the previously obtained interfaces. Testing can check Interface properties such as Activity components, Event Handlers, Widget Properties, and Widget Values. To support this extensive checking, original test cases can add specific assertions so that a test failure indicates an inconsistency in properties.

A tool named $A^2 T^2$ (Android Automatic Testing Tool) supports the proposed testing technique. $A^2 T^2$ includes a Java code instrumentation component, the GUI Crawler and a Test Case Generator based on the GUI tree. The GUI crawler uses Robotium [64], an Android testing framework providing facilities for the analysis of components in a running Android application. The Test Case Generator builds test cases as Java methods which replay event sequences and verify the presence of crashes (in crash testing) or the equivalence of user interface properties by means of assertions (in regression testing). Both the Crawler and the generated test cases can run in the context of the Android Emulator [8].

6.3 Random Testing Strategies

Liu et al. propose an approach for testing mobile applications that mixes elements of event-based testing and random testing [53]. The authors base their technique on a black-box model of the mobile application that considers just input and output at the application level. Sequences of GUI or context events represent the input of a mobile application. A context event may either be obtained from the physical context of the device (such as GPS receiver, Bluetooth chips, accelerometer sensor, humidity sensor, magnetic sensor, and network, etc.), or from social contexts (such as nearby chat friends, the current activity of a user, etc.). The mobile application must accept and react to both kinds of continuously changing context events as input to produce the correct output.

A pair $\langle t, v \rangle$ represents each event, where t represents the type of the event and v represents the value of the event. Therefore, sequences of event pairs form test cases, such that a test case T is defined as: $T = \langle e_1, e_2, \ldots, e_n \rangle$, where each $e_i = \langle t_i, v_i \rangle$.

Random test case generation follows from Adaptive Random Testing (ART) techniques. Researchers suggest that the ART approach that can offer increased effectiveness over purely random testing [38]. Random testing, indeed, has the benefits of low cost as a result of random generation of test cases and the ease of automation. Unfortunately, although simple and fully automatic, fully random techniques cannot offer guaranteed effectiveness in detecting faults.

ART, instead, builds on the observation that the input failure regions of an application tend to cluster together. Therefore, ART tries to spread the randomly generated test cases as evenly as possible using a measure of test case distance. Empirical studies [38] show ART's increased effectiveness over purely random testing.

Liu et al. apply the ART technique to test case generation for Android applications. A preliminary experimental study shows the ART technique to be more effective for test case generation than random testing in terms of fault detection in Android applications [53].

7. TESTING THE NON-FUNCTIONAL REQUIREMENTS OF MOBILE APPLICATIONS

Mobile applications, whether explicitly or implicitly stated, must usually satisfy a number of different non-functional requirements.

Common non-functional requirements include performance, compatibility, accessibility, usability, and security. Verifying each non-functional requirement requires the design of testing activities with specific goals.

In the following section, we discuss useful verification activities in the testing of the non-functional requirements of an Android mobile application.

7.1 Performance Testing

Performance refers to the degree to which an application completes its designated functions within given constraints [48]. In Android applications, performance testing aims to verify specified system performance goals, such as execution time, responsiveness, memory usage, power consumption, network usage, or additional goals.

Performance testing ultimately requires test execution in real environments by real devices to ensure the accuracy of performance measures. However, testers can also carry out performance testing in a simulated environment, both at system level and at unit level [50]. Of course, simulated performance tests in simulated environments only allow activities like the comparison of implementation alternatives, inefficiencies due to programming style, or the detection of performance bottlenecks early in the lifecycle. The Android community provides a set of techniques and advice for developing time- and space-efficient applications [13].

In the Android development environment, performance testing can exploit specific features provided by the Android platform and its testing utilities. As an example, the ACTION_BATTERY_CHANGED Intent message [16] monitors the battery usage of a device and the DDMS tool [26] can monitor device memory usage.

7.2 Stress Testing

Stress testing evaluates the behavior of the application when at or beyond any specified resource requirements [48].

The performance-sensitive stress tests always run on real devices in extreme execution environment conditions. The test aims to verify the system's tendency to crash and ability to recover from extreme conditions.

For example, the execution of multiple services on a device can slow down the execution of an application. Stress test cases can recreate very slow execution and check for time-out exceptions or application crashes. Stress tests can also verify the behavior of the application in scenarios of extreme usage or even exhaustion of resources such as available memory, disk space, or network bandwidth. The *Monkey* [24] automation tool can also generate and execute stress tests for applications by assessing their behavior in response to arbitrarily long event sequences.

7.3 Security Testing

Security testing aims to an application's ability to protect against undesired accesses by unauthorized users or processes and its ability to protect system resources from improper use while successfully granting access to authorized users, services, and resources.

Due to the unusual mix of personal and business uses typical of smartphone users, security represents a very important class of non-functional requirements for mobile applications. For a more complete overview of the types and natures of security threats and available security software and tools for security testing, consult Landman [52].

In the Android operating system, several protection mechanisms enforce security requirements. For example, each application runs a process with a unique Linux user ID and group ID, so that the application interacts with the Dalvik virtual machine as a separate user. This approach prevents processes from accessing resources external to their execution context.

Moreover, Android applications statically declare a set of permissions in the Android manifest file, which specifies the device features and operations that the application can access. The operating system provides a mechanism to inform the user about all the permissions requested by an application when installing the applications' APK on the device. Users can accept or reject installation after reviewing these permissions.

Lastly, developers must sign every application available on the Android market with a certificate and author-specific private key. The official Android Market of Google checks the signature of every application to ensure that only the original author of the application can release any upgrades [20]. Even applications running without the support of the Dalvik virtual machine and interacting directly interacting with Linux services must comply with the same policies.

The characteristics of the Android security model, with its main strengths, limitations, and problems, as well as techniques for making secure Android applications have been recently discussed [37, 42]. The research community also remains very interested in the security and privacy risks related to Android applications [31, 62, 63], and enthusiasts frequently post about these aspects of the Android framework in blogs and programming forums.

Researchers have recently discovered many security flaws in the platform In May 2011, researchers from the Ulm University discovered a security flaw involving most of the Android-powered devices which allowed impersonation attacks against Google services by exploiting the synchronization functionality of some Google applications included in the Android distributions (e.g., "Google Calendar" and "Google Contacts") [51]. Google quickly neutralized this flaw, but its presence emphasizes the importance of the security issues in the fast-growing Android world of quick device, system, and application releases.

Given the provided security mechanisms of the platform, a security test plan for an Android application should specify security requirements and tests to locate weaknesses or situations that would cause a violation of security requirements. Security testing ultimately requires planning outside of standard system tests, as the system must be secure against both unanticipated and anticipated attacks [72]. Some testers carry out security testing by inviting people to penetrate the system through security loopholes and application vulnerabilities.

As an example, in applications that access shared data, SQL injection exploits represent a well-known vulnerability caused when an application lacks input validation. Therefore, testers must write specific tests of the application's ability to sanitize inputs and protect against SQL injections. Testing activities should also consider other known patterns of attack and potential application-specific vulnerabilities.

7.4 Compatibility Testing

Compatibility testing aims to discover failures of the application due the variety in hardware and configuration profiles and failures related to use the application on non-targeted devices.

Since the Android world continues to grow very rapidly, vendors frequently release new features, devices, and software revisions (e.g., of the Android software platform and of third-party APIs). Because the market for Android application use includes all types of rapidly changing configurations, compatibility testing represents an important non-functional requirement.

Testers can partially execute cross-platform compatibility testing within emulated environments made up of virtual devices specified by various AVD files [17] to emulate the characteristics of different devices, hardware profiles, and operating system versions. The use of emulation drastically reduces the costs of compatibility testing by eliminating the need for a large number of physical devices for testing. Moreover, emulation provides the only alternative for yet-to-be-released devices, environments, and software platforms. However, because emulation is not always perfect, more reliable compatibility testing requires execution (when possible) in real environments provided by real devices.

Testers often perform compatibility testing within a well-defined reasonable scope. Compatibility testing does not need to check application behavior against every possible device configuration, but only configurations that the application specifically supports and any others relevant to the application's users (e.g., testing an application for a warning notice when using recent but unsupported platforms). The Android manifest documents the target operating system and APIs required by the application.

7.5 Usability Testing

The IEEE Glossary defines usability as "the ease with which a user can operate, prepare input for, and interpret output of a system or component" [48].

Mobile applications magnify the importance of usability requirements because the applications run on devices having small displays and offering heterogeneous styles of user interaction, such as touch, multi-touch, physical and virtual keyboards, location tracking, motion tracking, and so on [76].

As pointed out by Harty [46], many organizations do not carry out any usability or accessibility testing at all, considering these activities too expensive, too specialized, and less critical than functional testing. However, usability has proven itself as a fundamental factor in the success and increase of Android applications and Android-powered devices.

The Android Developer reference guide [25] proposes guidelines for usability in Android applications. The guidelines propose techniques for designing usable AppWidgets, Activities, and Menus.

Because device capabilities vary, testers should pursue usability specification in conjunction with compatibility, in order to assure usability on all target devices and configurations. To this aim, some researchers have proposed guidelines and mechanisms for systematically defining auto-adaptive Activity layouts supporting usability testing by considering all screen configurations [22].

Android applications have a particular sensitivity to two particular aspects of usability responsiveness and seamlessness.

We define responsiveness as the capability of the system to react readily to user input. Android applications can suffer from responsiveness problems due to the existence of concurrent background services and activities that may slow processor performance. As a result, the user must interact with a very slow, unresponsive interface [14].

We define seamlessness as the ability of the system to avoid situations in which the user experience suffers from frequent interruptions in Activity execution. Notifications and messages coming from background Service and Broadcast Receiver components can cause these interruptions. The current Android platform remains sensitive to this problem in part because the inability to split the device screen into separate windows means that user notifications and dialogs must remain in the foreground [15].

Usability testing usually involves a large set of users of the application and an analysis of their interaction with the application. Testing of this magnitude presents two important problems, including the high costs and effort needed to execute the tests and obtain feedback and the inability to execute any test prior to the release of an alpha version of the application. Automation can reduce the cost and effort of usability testing. Testing activities may also begin throughout system testing rather than waiting for a release on a real physical device.

In the realm of current research on mobile device usability testing, Au et al. present some techniques, tools, and experiments for the usability testing of handheld applications [28]. These approaches apply to the Android platform because they focus on similar characteristics and limitations of the device, such as screen capabilities. Harty also addresses some usability aspects such as keyboard navigation and layout issues [46].

7.6 Accessibility Testing

Accessibility testing represents a particular critical type of usability testing aimed at the verification of application performance for users with physical disabilities. The Android system offers a general accessibility layer providing features such as text-to-speech, haptic feedback, and trackball navigation that augment the user experience as needed for users with physical disabilities.

Recent research in the area of accessibility services offers additional approaches for accessibility testing. The TalkBack accessibility service by the Eyes-Free Project [75] allows use of the device as a blind user by providing support for blanking the screen, allowing only spoken feedback, and using enabling only the directional controller instead of the touch controls.

8. TOOLS AND FRAMEWORKS FOR ANDROID TESTING AUTOMATION

Given the needs presented in the previous sections for both functional and non-functional testing of Android applications, we now discuss the role of automation in these testing activities. Tools and technologies for software testing automation can significantly contribute to the effectiveness of a testing process. In general, testing tools can automate some of the tasks of the testing process, such as test case generation, test case execution, and evaluation of test case results. Moreover, testing tools may support the production of useful testing documentation and provide a configuration management of this documentation alongside the application source code.

Depending on the testing activity they support, two main categories of tools can be considered: tools for functional requirements testing and tools for non-functional requirement testing, such as performance, load, security, or usability testing.

In recent years, researchers have developed several tools and technologies to aid the testing of mobile applications. To-date, however, most of these tools have been developed for mobile platforms other than Android. Existing tools frequently offer the following features:

- *Performing Black-Box Testing:* like the features offered by MobileTest proposed by Bo et al. [34], or by Hermes, a tool for description and automatic execution of test cases for Java applications developed to run in J2ME environments [70]. Test cases in Hermes are written in XML and can be executed automatically on the actual device, rather than on device emulators through a distributed run-time environment.
- *Performing White-Box Coverage Testing:* like the features implemented by JaBUTi/ME, a tool providing an environment for executing coverage testing of J2ME applications not only on emulators, but also on their real target mobile devices with the aid of structural coverage assessment [39].
- *Performing Performance Testing:* like the JUnit-based tool presented by Kim et al. [50] which supports performance testing at the unit level of mobile applications implemented in the J2ME environment.
- *Implementing Cross-Network Testing of Mobile Applications:* like the application-level emulator for mobile computing devices proposed by Satoh [66, 67] that enables application-level software execution and testing in mobility, as supported by the services and resources provided through its current network environment.

As to mobile applications developed for the Android platform, the Android development platform provides most resources and technologies for automation [19]. The Android SDK builds on a Java SDK containing all Java code libraries needed to create applications that run specifically on the Android platform. The SDK also includes help files, documentation, the

Android Emulator, and other tools for development and debugging, such as the previously described *adb* [7] and *DDMS* [26] tools.

The Open Handset Alliance has released an Android plugin for Eclipse, the ADT [6] (Android Development Tools). This plugin integrates the creation, compilation, packaging, exporting, testing, debugging, and even the emulation of Android-specific projects in the familiar Eclipse environment.

All the testing activities for Android likely exploit the Android Testing framework, a rich set of APIs based on JUnit and extended with an instrumentation framework and Android-specific testing classes. The instrumentation framework contains specific classes for running test cases on Android applications (in particular, these classes provide hooks to control any Android component independently of its normal lifecycle). The Android-specific testing classes extend the JUnit *TestCase* and *Assert* with Android-specific setup, teardown, and helper methods and classes. The framework also provides component-specific test case classes, such as classes for testing Activity, Content Provider, and Service components directly. Moreover, the framework includes specific classes for testing components in isolation from the rest of the system, which provides excellent support for system and integration levels of testing. These classes include Mock classes that can isolate tests from the rest of the system and facilitate dependency injection by stubbing out or overriding normal operations and Context classes for setting specific contexts for test execution.

The SDK also provides two tools for functional application testing:

- The UI/Application Exerciser *Monkey* [24], usually called "monkey", is a command-line tool that sends pseudo-random streams of keystrokes, touches, and gestures to a device. *Monkey* runs with the Android Debug Bridge (adb) tool, and can stress test an application and report any encountered errors. The tool can also repeat a stream of events if run with the same random number seed.

- The *monkeyrunner* tool [18] provides an API and an execution environment for the testing of Android applications from test programs written in Python. The API includes functions for connecting to a device, installing and uninstalling packages, taking screenshots, comparing two images, and running a test package against an application. The flexibility of *monkeyrunner* makes a wide range of automated testing activities possible.

An additional testing tool called Robotium is available at the project's Google Code site [64]. This test framework aims to make writing powerful and robust automatic black-box test cases for Android applications easy.

Robotium's approach for deriving test cases compares to capture and replay tools like the Web testing framework Selenium [68]. With the support of Robotium, test case developers can write function, system and acceptance test scenarios, spanning multiple Android Activity components.

The Android testing framework and tools described above provide basic support for automating testing activities in any type of testing process. So far, these tools have been used to implement other prototype tools, like the TEMA tools [74] for model-based testing that we described in Section 6 and the A^2T^2 for GUI-based testing of Android applications [4].

In the future, additional tools will potentially make Android testing automation feasible and effective. As an example, tools for functional requirement testing of Android applications should provide features for automating the following basic testing activities:

- *Test Model Generation:* to produce an instance of the desired test model of the subject application. This model may be a model already produced along the development process (so that the tool would just have to import it), or it may be produced by reverse engineering techniques.
- *Code Instrumentation:* to instrument automatically the code of an Android application by inserting utilities for automatically collecting data about test case execution.
- *Driver and Stub Generation:* to help produce the code of Android application components needed for test case execution, such as driver, stub, and mock modules.
- *Test Case Management:* to automatically generate test cases and to support test case design and testing documentation management.
- *Test Case Execution:* to support the automatic execution and evaluation of test cases on the Android application under test.
- *Test Result Analysis:* to analyze and to automatically evaluate test case results.
- *Report Generation:* to produce adequate reports analyzing test results, such as coverage reports about the components exercised during the test.

Here, we only list tools for functional requirements testing activities, but non-functional requirements testing activities also need adequate tool support.

9. CONCLUDING REMARKS

The growing request for quality mobile applications continues to drive research interest in mobile application testing. The dominant role assumed by applications developed for open mobile platforms makes Android applications a reasonable target for this research. Several initiatives currently address the problem of finding testing principles, guidelines, models, techniques, and technologies that enable more effective and efficient testing processes for these applications.

In this chapter, we analyzed the main challenges and open issues characterizing the field of mobile application testing. We presented the advances and their ramifications in the specific context of Android application testing.

We analyzed how the same basic principles, methods, and techniques that apply to any software testing process can be adapted to Android mobile application testing. Consequently, we discussed feasible solutions for carrying out Android application testing at unit, integration, and system levels. We examined some strategies for Android testing recently proposed in research literature. We also discussed principles for carrying out testing of non-functional requirements of Android applications. Finally, we reported the state of the art of technologies and tools for automating testing activities in real processes.

We conclude primarily that while some relevant preliminary contributions exist, other topics have only marginally been addressed to-date. Aspects of Android testing that need further investigation and emphasis in the future include:

- Finding suitable models for representing the Android systems or components under test.
- Defining effective techniques for automatically deriving test cases from these models.
- Developing suitable strategies for managing test suites and minimizing their size.
- Developing specific tools for aiding the most expensive steps of Android testing.
- Empirical validations of Android testing procedures, including these proposals.

REFERENCES

[1] D. Amalfitano, A.R. Fasolino, P. Tramontana, Reverse engineering finite state machines from rich internet applications, in: Proceedings of the Working Conference on Reverse Engineering (WCRE 2008), 2008, IEEE Computer Society Press, pp. 69–73.

[2] D. Amalfitano, A.R. Fasolino, P. Tramontana, Rich internet application testing using execution trace data, in: Proceedings of Second International Workshop on TESTing Techniques and Experimentation Benchmarks for Event-Driven Software (TESTBEDS 2010), IEEE Computer Society Press, pp. 274–283.

[3] D. Amalfitano, A.R. Fasolino, P. Tramontana, Techniques and tools for rich internet applications testing, in: Proceedings of the 12th Symposium on Web Systems Evolution (WSE 2010), IEEE Computer Society Press, pp. 63–72.

[4] D. Amalfitano, A.R. Fasolino, P. Tramontana, A GUI crawling-based technique for android mobile application testing, in: Third International Workshop on TESTing Techniques and Experimentation Benchmarks for Event-Driven Software (TESTBEDS 2011), IEEE Computer Society Press, pp. 252–261.

[5] Android Developers, The Developer's Guide, Activities. <http://developer.android.com/guide/topics/fundamentals/activities.html> (accessed 30.07.11).

[6] Android Developers, The Developer's Guide, ADT Plugins for Eclipse. <http://developer.android.com/sdk/eclipse-adt.html> (accessed 30.07.11).

[7] Android Developers, The Developer's Guide, Android Debug Bridge. <http://developer.android.com/guide/developing/tools/adb.html> (accessed on 30.07.11).

[8] Android Developer, The Developer's Guide, Android Device Emulator. <http://developer.android.com/guide/developing/tools/emulator.html> (accessed on 30.07.11).

[9] Android Developers. The Developer's Guide. Android Virtual Device (AVD). <http://developer.android.com/guide/developing/devices/index.html> (accessed on 30.07.11).

[10] Android Developers, The Developer's Guide, Application Fundamentals. <http://developer.android.com> (accessed on 30.07.11).

[11] Android Developers, The Developer's Guide, Content Providers. <http://developer.android.com/guide/topics/providers/content-providers.html> (accessed on 30.07.11).

[12] Android Developers, The Developer's Guide, Context. <http://developer.android.com/reference/android/content/Context.html> (accessed on 30.07.11).

[13] Android Developers, The Developer's Guide, Designing for Performance. <http://developer.android.com/guide/practices/design/performance.html> (accessed on 30.07.11).

[14] Android Developers, The Developer's Guide, Designing for Responsiveness. <http://developer.android.com/guide/practices/design/responsiveness.html> (accessed on 30.07.11).

[15] Android Developers, The Developer's Guide, Designing for Seamlessness. <http://developer.android.com/guide/practices/design/seamlessness.html> (accessed on 30.07.11).

[16] Android Developers, The Developer's Guide, Designing for Seamlessness. <http://developer.android.com/reference/android/content/Intent.html> (accessed on 30.07.11).

[17] Android Developers, The Developer's Guide, Managing Virtual Devices. <http://developer.android.com/guide/developing/devices/index.html> (accessed on 30.07.11).

[18] Android Developers. The Developer's Guide, Monkeyrunner. <http://developer.android.com/guide/developing/tools/monkeyrunner_concepts.html> (accessed on 30.07.11).

[19] Android Developers, The Developer's Guide, Android SDK. <http://developer.android.com/sdk/index.html> (accessed on 30.07.11).

[20] Android Developers, The Developer's Guide, Android Security and Permissions. <http://developer.android.com/guide/topics/security/security.html> (accessed on 30.07.11).

[21] Android Developers, The Developer's Guide, Services. <http://developer.android.com/guide/topics/fundamentals/services.html> (accessed on 30.07.11).

[22] Developers, The Developer's Guide, Supporting Multiple Screens. <http://developer.android.com/guide/practices/screens_support.html> (accessed on 30.07.11).

[23] Android Developers, The Developer's Guide, Testing. <http://developer.android.com/guide/topics/testing/index.html> (accessed on 30.07.11).

[24] Android Developers, The Developer's Guide, UI/Application Exerciser Monkey. <http://developer.android.com/guide/developing/tools/monkey.html> (accessed on 30.07.11).

[25] Android Developer, The Developer's Guide, User Interface Guidelines. <http://develo per.android.com/guide/practices/ui_guidelines/index.html> (accessed on 30.07.11).

[26] Android Developers, The Developer's Guide, Using DDMS. <http://developer.android. com/guide/developing/debugging/ddms.html> (accessed on 30.07.11).

[27] Android Developers, The Developer's Guide, What is Android? <http://developer. android.com/guide/basics/what-is-android.html> (accessed on 30.07.11).

[28] Fiora T.W. Au, Simon Baker, Ian Warren, Gillian Dobbie, Automated usability testing framework, in: Proceedings of the ninth conference on Australasian user interface (AUIC '08), 2008, Vol. 76, Australian Computer Society, Inc., Darlinghurst, Australia, pp. 55–64.

[29] Apple App Store. <http://www.apple.com/iphone/apps-for-iphone/> (accessed on 30.07.11).

[30] Guardian.co.uk, Apple's iOS App Store reaches 15bn downloads milestone. <http:// www.guardian.co.uk/technology/appsblog/2011/jul/07/apple-iphone-app-store-downloads> (accessed on 30.07.11).

[31] David Barrera, H. Günes Kayacik, Paul C. van Oorschot, Anil Somayaji, A methodol-ogy for empirical analysis of permission-based security models and its application to android, in: Proceedings of the 17th ACM conference on Computer and communica-tions security (CCS '10), 2010 CM, New York, NY, USA, pp. 73–84.

[32] F. Belli, C.J. Budnik1, L. White, Event-based modelling, analysis and testing of user interactions: approach and case study, Softw. Test. Verif. Reliab. (16) (2006) 3–32.

[33] R.V. Binder, Testing Object-Oriented Systems- Models, Patterns, and Tools, Addison-Wesley, Boston, MA, USA, 1999.

[34] J. Bo, L. Xiang, G. Xiaopeng, Mobiletest, A Tool Supporting Automatic Black Box Testing for Software on Smart Mobile Devices, in: AST '07: Proceedings of the Sec-ond International Workshop on Automation of Software Test, IEEE Computer Society, Washington, DC, USA, 2007, pp. 8–14.

[35] R.C. Bryce, S. Sampath, A.M. Memon, Developing a single model and test prioritization strategies for event-driven software, IEEE Trans. Softw. Eng. 37 (1) (2011) 48–64.

[36] Bart Broekman, Edwin Notenboom, Testing embedded software, Pearson Education, 2003.

[37] Jesse Burns, Developing Secure Mobile Applications for Android, ISEC Partners. <http://www.isecpartners.com/files/iSEC_Securing_Android_Apps.pdf> (accessed on 30.07.11).

[38] T.Y. Chen, F.-C. Kuo, R.G. Merkel, T.H. Tse, Adaptive random testing: the ART of test case diversity, J. Sys. Softw. (2009). <http://dx.doi.org/10.1016/j.jss.2009.02.022>.

[39] M.E. Delamaro, A.M.R. Vincenzi, J.C. Maldonado, A strategy to perform coverage testing of mobile applications, in: Proceedings of the 2006 international workshop on Automation of software test (AST '06), ACM, New York, NY, USA, pp. 118–124.

[40] J.F. Di Marzio, Android- A Programmer's Guide, McGraw Hill, 2008.

[41] M. Doernhoefer, Software Engineering for Mobile Devices, ACM SIGSOFT Software Engineering Notes 35 (5) (2010) 8–17.

[42] W. Enck, M. Ongtang, P. McDaniel, Understanding android security and privacy, IEEE 7 (1) (2009) 50–57.

[43] B. Fling, Mobile Design and Development, O' Reilly, 2009.

[44] Gartner Newsroom, Gartner Says Android to Become No. 2 Worldwide Mobile Operat-ing System in 2010 and Challenge Symbian for No. 1 Position by 2014. <http://www.gartner.com/it/page.jsp?id=1434613> (accessed on 30.07.11).

[45] D. Gavalas, D. Economou, Development platforms for mobile applications: status and trends, IEEE Software 28 (1) (2011) 77–86.

[46] Julian Harty, Finding usability bugs with automated tests, Queue 9 (1) (2011) 20–27.

[47] C. Hu, I. Neamtiu, Automating GUI Testing for Android Applications, in: Proceedings of AST 2011, 6th International Workshop on Automation of Software Test, ACM Press, pp. 77–83.

[48] IEEE Std. 610.12- 1990, Glossary of Software Engineering Terminology, in: Software Engineering Standard Collection, IEEE CS Press, Los Alamitos, California, 1990.

[49] Junit, Resources for test driven development. <http://www.junit.org> (accessed on 30.07.11).

[50] H. Kim, B. Choi, W. Eric Wong, Performance testing of mobile applications at the unit test level, in: Proceedings of 2009 Third IEEE International Conference on Secure Software Integration and Reliability Improvement, IEEE Comp. Soc. Press, 171–181.

[51] Bastian Könings, Jens Nickels, Florian Schaub, Catching authTokens in the wild the insecurity of google's clientlogin protocol. <http://www.uni-ulm.de/en/in/mi/staff/koenings/catching-authtokens.html> (accessed on 30.07.11)

[52] Max Landman, Managing smart phone security risks, in: 2010 Information Security Curriculum Development Conference (InfoSecCD '10), ACM, New York, NY, USA, 2010, pp. 145–155.

[53] Z. Liu, X. Gao, Xiang Long, Adaptive random testing of mobile Application, in: 2nd International Conference on Computer Engineering and Technology (ICCET), 2, 2010 297–301.

[54] Q.H. Mahmoud, Testing wireless java applications, Sun Microsystems—on-line article. <http://www.oracle.com/technetwork/systems/test-156866.html>, 2002 (accessed on 30.07.11).

[55] A.K. Maji, K. Hao, S. Sultana, S. Bagchi, Characterizing failures in mobile OSes: a case study with android and symbian, in: IEEE 21st International Symposium on Software Reliability Engineering, IEEE Computer Society Press, 2010, pp. 249–258.

[56] A. Marchetto, P. Tonella, F. Ricca, State-based testing of ajax web applications, in: Proceedings of 2008 International Conference on Software Testing, Verification and Validation, IEEE CS Press, 2008, pp. 120–130.

[57] A. Memon, L. Banerjee, A. Nagarajan, GUI ripping: reverse engineering of graphical user interfaces for testing, in: Proceedings of the 10th Working Conference on Reverse Engineering (WCRE 2003), 2003, IEEE Computer Society Press, pp. 260–269.

[58] Atif Memon, Qing Xie, Using transient/persistent errors to develop automated test oracles for event-driven software, in: 19th IEEE International Conference on Automated Software Engineering (ASE'04), IEEE Computer Society Press, 2004, 186–195.

[59] A.M. Memon, Q. Xie, Studying the fault-detection effectiveness of GUI test cases for rapidly evolving software, IEEE Trans. Softw. Eng. 31 (10) (2005) 884–896.

[60] A.M. Memon, Q. Xie, Designing and comparing automated test oracles for GUI-based software applications, ACM Transactions Software Engineering and Methodology ACM Press 16 (1) (2007).

[61] A. Mesbah, A. van Deursen, Invariant-based automatic testing of AJAX user interfaces, in: Proceedings of International Conference on Software Engineering (ICSE 2009), IEEE Computer Society Press, 2009 pp. 210–220.

[62] Mohammad Nauman, Sohail Khan, Xinwen Zhang, Apex: extending Android permission model and enforcement with user-defined runtime constraints, in: Proceedings of the 5th ACM Symposium on Information, Computer and Communications Security (ASIACCS '10), ACM, New York, NY, USA, 2010, pp. 328-332.

[63] Machigar Ongtang, Kevin Butler, Patrick McDaniel, Porscha: policy oriented secure content handling in android, in: Proceedings of the 26th Annual Computer Security Applications Conference (ACSAC '10), ACM, New York, NY, USA, 2010, pp. 221–230.

[64] Robotium. <http://code.google.com/p/robotium/> (accessed on 30.07.11).

[65] BBC news online, Google bets on android future. <http://news.bbc.co.uk/2/hi/7266201.stm> (accessed on 30.07.11).

[66] I. Satoh, A testing framework for mobile computing software, IEEE Trans. Softw. Eng. 29 (12) (2003) 1112–1121.

[67] I. Satoh, Software testing for wireless mobile application, IEEE Wireless Communications, (2004) 58–64.

[68] SeleniumHQ, Web application testing system. <http://seleniumhq.org/> (accessed on 30.07.11).

[69] R.K. Shehady, D.P. Siewiorek, A method to automate user interface testing using variable finite state machines, in: Proceedings of the 27th International Symposium on Fault-Tolerant Computing (FTCS '97), IEEE Computer Society, Washington, DC, USA, pp. 80–88.

[70] S. She, S. Sivapalan, I. Warren, Hermes: a tool for testing mobile device applications, in: Proceedings of 2009 Australian Software Engineering Conference, IEEE Comp. Soc. Press, pp. 123–130.

[71] Sensor Simulator. <http://code.google.com/p/openintents/wiki/SensorSimulator> (accessed on 30.07.11).

[72] Ian Sommerville, Software Engineering, eigth ed., Pearson Education, 2006.

[73] S. Srirama, R. Kakumani, A. Aggarwal, P. Pawar, Effective testing principles for the mobile data services applications, in: First International Conference on Communication System Software and Middleware, Comsware, IEEE Computer Society Press, 2006, pp. 1–5.

[74] T. Takala, M. Katara, J. Harty, Experiences of system-level model-based GUI testing of an android application, in: Fourth IEEE International Conference on Software Testing, Verification and Validation, IEEE Computer Society Press, 2011, pp. 377–386.

[75] TalkBack, Eyes-free project. <https://market.android.com/details?id=com.google.android.marvin.talkback> (accessed on 30.07.11).

[76] A. Wasserman, Software engineering issues for mobile application development, in: Proceedings of the FSE/SDP Workshop on Future of Software Engineering Research, FOSER 2010, IEEE Computer Society Press, pp. 397–400.

[77] J.L. Wesson, D.F. van der Walt, Implementing mobile services: does the platform really make a difference? in: SAICSIT '05: Proceedings of the 2005 Annual Research Conference of the South African Institute of Computer Scientists and Information Technologists on IT Research in Developing Countries, South African Institute for Computer Scientists and Information Technologists, 2005, pp. 208–216.

[78] L. White, H. Almezen, Generating test cases for GUI responsibilities using complete interaction sequences, in: Proceedings of the 11th International Symposium on Software Reliability Engineering (ISSRE '00), IEEE Computer Society, Washington, DC, USA, 2000, pp. 110–121.

[79] Wikipedia, List of digital distribution platforms for mobile devices. <http://en.wikipedia.org/wiki/List_of_digital_distribution_platforms_for_mobile_devices> (accessed on 30.07.11).

[80] X. Yuan, A.M. Memon, Generating event sequence-based test cases using GUI run-time state, IEEE Trans. Softw. Eng. 36 (1) (2010) 81–95.

ABOUT THE AUTHORS

Domenico Amalfitano received the Laurea degree in Computer Engineering and the Ph.D. degree in Computer Engineering and Automation in 2011 by the University of Naples "Federico II". Currently he is a postdoctoral researcher at the University of Naples "Federico II" and his research interests concern the topics of Software Engineering, mainly including the reverse engineering, comprehension, migration, testing and testing automation of Event

Driven Software Systems, mostly in the fields of Web Applications, Mobile Applications and GUIs. Now, he is also applying his research experience in the Italian automotive industry.

Anna Rita Fasolino received the Laurea degree in Electronic Engineering in 1992 and a Ph.D. in Electronic and Computer Engineering in 1996 by the University of Naples "Federico II". In 1998 she was an Assistant Professor in Computer Science at the University of Bari and in 1999 she moved to the Faculty of Engineering of the University Federico II in Naples. Since January 2005 she is an Associate Professor at the University of Naples. Professor Fasolino's main research interests are in the field of Software Engineering and include Software Maintenance, Reverse Engineering, Web Engineering, Software Testing and Quality. In these areas she published more than 70 papers in international scientific journals, books, and proceedings of international Conferences. She participates to the Program Committee of several international Conferences in the software engineering field and acts as a reviewer in Journals and International research projects. Professor Fasolino takes part to and leads the research activities in National and International Research Projects on Software Engineering topics in collaboration with Italian Universities and Industries.

Porfirio Tramontana received the Laurea degree in Computer Engineering and a Ph.D. degree in Computer Engineering in 2005. He is currently Assistant Professor at the University of Naples "Federico II". His main research interests regard the software engineering field. They include testing, reverse engineering, migration, quality assessment and semantic interoperability. In these research fields, he has published about 40 papers appeared in international journals, books and conference proceedings.

Bryan Robbins holds a bachelor's degree in software engineering and a master's degree in computer science, both from Mississippi State University. He is currently pursuing his doctoral work in computer science at the University of Maryland - College Park. His current research focuses on model-based GUI testing and test automation with the GUITAR framework (http://guitar.sourceforge.net). He also works in the financial services industry as a automation and quality assurance specialist for enterprise web applications.

CHAPTER TWO

Regression Testing of Evolving Programs

Marcel Böhme, Abhik Roychoudhury, and Bruno C.d.S. Oliveira
National University of Singapore, Singapore

Contents

Advances in Computers, Volume 89
ISSN 0065-2458, http://dx.doi.org/10.1016/B978-0-12-408094-2.00002-3

Abstract

Software changes, such as bug fixes or feature additions, can introduce software bugs and reduce the code quality. As a result tests which passed earlier may not pass any more—thereby exposing a regression in software behavior. This survey overviews recent advances in determining the impact of the code changes onto the program's behavior and other syntactic program artifacts. Static program analysis can help determining change impact in an approximate manner while dynamic analysis determines change impact more precisely but requires a regression test suite. Moreover, as the program is changed, the corresponding test suite may, too. Some tests may become obsolete while other tests are to be augmented that stress the changes. This article surveys such test generation techniques to stress and propagate program changes. It concludes that a combination of dependency analysis and lightweight symbolic execution show promise in providing powerful techniques for regression test generation.

1. INTRODUCTION

Software Maintenance is an integral part of the development cycle of a program. In fact, the evolution and maintenance of a program is said to account for 90% of the total cost of a software project, prompting the authors to call it the legacy crisis [1]. The validation of such ever-growing, complex software programs becomes more and more difficult. Manually generated test suites increase in complexity as well. In practice, programmers tend to write test cases only for corner cases or to satisfy specific code coverage criteria.

Regression testing builds on the assumption that an existing test suite stresses much of the behavior of the existing program P implying that at least one test case fails upon execution on the modified program P' when P is changed and its behavior regresses [2]. Informally, if the developer is confident about the correctness of P, she has to check only whether the changes introduced any regression errors in order to assess the correctness of P'. This implies that the testing of evolving programs can focus primarily on the syntactic (and semantic) entities of the program that are affected by the syntactic changes from one version to the next.

The importance of automatic regression testing strategies is unequivocally increasing. Software regresses when existing functionality stops working upon the change of the program. A recent study [3] suggests that even intended code quality improvements, such as the fixing of bugs, introduces new bugs in 9% of the cases. In fact, at least 14.8–24.4% of the security patches released by Microsoft over 10 years are incorrect [4].

The purpose of this chapter is to provide a survey on the state-of-the-art research in testing of evolving programs. This chapter is structured as follows.

In Section 2, we present a quick overview of dependency analysis and symbolic execution which can help to determine whether the execution and evaluation of one statement influences the execution and evaluation of another statement. In particular, we discuss program slicing as establishing the relationship between a set of syntactic program elements and units of program behavior. In Section 3 we survey the related work of change impact analysis which seeks to reveal the syntactic program elements that may be affected by the changes. In particular, we discuss the problem of semantic change interference, for which the change of one statement may semantically interfere or interact with the change of another statement on some input but not on others. These changes cannot be tested in isolation. Section 4 highlights the salient concepts of regression testing. We show that the adequacy of regression test suites can be assessed in terms of code coverage which may approximate the measure of covered program behavior. For instance, a test suite that is 95% statement coverage-adequate exercises exactly 95% of the statements in a program. Section 5 investigates the *removal* of test cases from an existing test suite that are considered *irrelevant in some respect*. In many cases, a test case represents an equivalence class of input with similar properties. If two test cases represent the same equivalence class, one can be removed without reducing the current measure of adequacy. For instance, a test case in a test suite that is 95% statement coverage-adequate may represent a certain statement and thus every input exercising that statement. We may be able to remove a few test cases from that test suite without decreasing the coverage below 95%. Similarly, Section 6 investigates the *augmentation* of test cases to an existing test suite that are considered *relevant in some respect*. If there is an equivalence class that is not represented, a test case may be added that represents this equivalence class. In the context of evolving programs it may be of interest to generate test cases that expose the behavioral difference introduced by the changes. Only difference-revealing test cases can expose software regression.

2. PRELIMINARIES

Dependency analysis and symbolic execution can help to determine whether the execution and evaluation of a statement s_1 influences the execution and evaluation of another statement s_2. In theory, it is generally undecidable whether there exists a feasible path (exercised by a concrete program input) that contains instances of both statements [5]. Static program analysis can approximate the potential existence of such paths for which both statements are executed and one statement "impacts" the other. Yet, this includes infeasible ones. Symbolic execution (SE) facilitates the exploration of all feasible program paths if the exploration terminates. In practice, SE allows to search for input that exercises a path that contains both statements.

2.1 Running Example

The program P on the left-hand side of Fig. 1 takes values for the variables i and j as input to compute output o. Program P is changed in three locations to yield the modified program version P' on the right-hand side. Change $ch1$ in line 2 is exercised by every input while the other two changes are guarded by the conditional statements in lines 5 and 9. Every change assigns the old value plus one to the respective variable. In this survey, we investigate which program elements are affected by the changes, whether they can be tested in isolation, and how to generate test cases that witness the "semantic impact" of these changes onto the program. In other words, in order to test whether the changes introduce any regression errors, we explain how to generate program input that produces different output upon execution on both versions.

2.2 Program Dependence Analysis

Static program analysis [6, 7] can approximate the "impact" of s_1 onto s_2. In particular, it can determine that there *does not exist* an input so that the execution and value of s_2 depends on the execution and value of s_1. Otherwise, static analysis can only suggest that there may or may not be such an input.

Statement s_2 *statically control-depends* on s_1 if s_1 is a conditional statement and can influence whether s_2 is executed [7]. Statement s_2 *statically*

```
1   input(i,j);
2   a = i;          //ch1 (a=i+1)
3   b = 0;
4   o = 0;
5   if(a > 0){
6       b = j;      //ch2 (b=j+1)
7       o = 1;
8   }
9   if(b > 0)
10      o = 2;       //ch3 (o=o+1)
11  output(o);
```

```
1   input(i,j);
2   a = i + 1;      //ch1 (a=i)
3   b = 0;
4   o = 0;
5   if(a > 0){
6       b = j + 1;  //ch2 (b=j)
7       o = 1;
8   }
9   if(b > 0)
10      o = o + 1;   //ch3 (o=2)
11  output(o);
```

Original Version P Modified Version P'

Fig. 1. Running example.

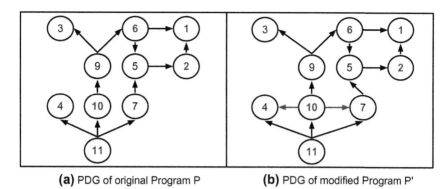

(a) PDG of original Program P   **(b)** PDG of modified Program P'

Fig. 2. Program dependency graph of running example.

data-depends on s_1 if there is a sequence of variable assignments[1] that potentially propagate data from s_1 to s_2 [7]. The *Control-Flow Graph* (CFG) models the static control dependence for every statement in the program. The statements are represented as nodes. Arcs pointing away from a node represent possible transfers of control to subsequent nodes. A program's entry and exit points are represented by initial and final vertices. So, a program can potentially be executed along paths leading from an initial to a final vertice. The *Def/Use Graph* extends the CFG and labels every node n by the variables defined and used in n. Another representation of the dependence relationship among the statements in a program is the Program Dependence Graph (PDG) [8]. Every statement s_2 is a node that has an outgoing arc to another statement s_1 if s_2 directly (not transitively) data- or control-depends on s_1. A statement s_2 *syntactically depends* on s_1 if in the PDG s_1 is reachable from s_2.

The program dependence graphs for both versions of the program in our running example are depicted in Fig. 2. The nodes are labeled by the line number. The graph is directed as represented by the arrows pointing from one node to the next. It does not distinguish data- or control dependence. For instance, the node number 7 transitively data- or control-depends on the node number 1 but not on nodes number 6 or 3 in both versions.

2.3 Program Slicing

A *program slice* of a program P is a reduced, executable subset of P that computes the same function as P does in a subset of variables at a certain point of interest, referred to as *slicing criterion* [9–12].

---

[1]A variable defined earlier is used later in the sequence.

A *static backward slice* of a statement *s* contains all program statements that potentially contribute in computing *s*. Technically, it contains all statements on which *s* syntactically depends, starting from the program entry to *s*. The backward slice can be used in debugging to find all statements that influence the (unanticipated) value of a variable in a certain program location. For example, the static backward slice of the statement in line 6 includes the statements in lines 1, 2, and 5. Similarly, a *static forward slice* of a statement *s* contains all program statements that are potentially "influenced" by *s*. Technically, it contains all statements that syntactically depend on *s*, starting from *s* to every program exit. A forward slice reveals which information can flow to the output. It might be a security concern if confidential information is visible at the output. As shown in Fig. 3, for our running example, the static forward slice of the statement in line 6 includes the statements in lines 9, 10, and 11.

If two static program slices are isomorphic, they are behaviorally equivalent [13]. In other words, if every element in one slice corresponds to one element in the other slice, then the programs constituted on both slices compute the same output for the same input. Static slices can be efficiently computed using the PDG (or System Dependence Graph (SDG)) [8, 10]. It is possible to test the isomorphism of two slices in linear time [12].

However, while a static slice considers all potential, terminating executions, including infeasible ones, a *dynamic slice* is computed for a given (feasible) execution [11]. A dynamic backward slice can resolve much more precisely which statements (actively) contribute in computing a given slicing criterion. Dynamic slices are computed based on the execution trace of an input. An execution trace contains the sequence of statement instances exercised by the input. In other words, input exercising the same program path produces the same execution trace.

The *relevant slice* for a slicing criterion s_i is the dynamic backward slice of s_i augmented by potential dependencies and contains all statement instances

Line	Type	Slice
2	Forward	2, 5, 6, 7, 9, 10, 11
	Backward	1
6	Forward	6, 9, 10, 11
	Backward	1, 2, 5, 6
10	Forward	10, 11
	Backward	1, 2, 3, 5, 6, 9, 10

Original Version *P*

Line	Type	Slice
2	Forward	2, 5, 6, 7, 9, 10, 11
	Backward	1
6	Forward	6, 9, 10, 11
	Backward	1, 2, 5, 6
10	Forward	10, 11
	Backward	1, 2, 3, 5, 6, **7**, 9, 10

Modified Version *P'*

Fig. 3. Static backward and forward slices.

in the execution trace that contribute (actively and passively) in computing s_i [14]. More specifically, every input exercising the same relevant slice computes the same symbolic values for the variables used in the slicing criterion [15]. For instance, in the context of debugging the developer might be interested in only those statements that led to the (undesired) value of the variable at a given statement for that particular, failing execution. Furthermore, relevant slices can be utilized for the computation of program summaries. By computing relevant slices w.r.t. the program's output statement, we can derive the symbolic output for a given input. Using path exploration based on symbolic output, we can gradually reveal the transformation function of the analyzed program and group input that computes the same symbolic output [15].

2.4 Symbolic Execution

While static analysis may suggest the potential existence of a path that exercises both statements so that one statement influences the other statement, the path may be infeasible. In contrast, *Symbolic Execution* (SE) [16–18] facilitates the exploration of feasible paths by generating input that each exercises a different path. If the exploration terminates, it can guarantee that there exists (or does not exist) a feasible path and program input, respectively, that exercises both statements. The test generation can be directed towards executing s_1 and s_2 in a goal-oriented manner [19–22].

SE generates for each test input a condition as first-order logic formula that is satisfied by every input exercising the same program path. This *path condition* is composed of a branch condition for each exercised conditional statement instance (e.g., `If` or `While`). A conjunction of branch conditions is satisfied by every input evaluating the corresponding conditional statements in the same direction. The negation of these branch conditions one at a time, starting from the last, allows to generate input that exercises the "neighboring" paths. This procedure is called *path exploration*.

The symbolic execution of our running example can reveal the symbolic program summaries in Fig. 4. Both versions have two conditional statements. So there are potentially $2^2 = 4$ paths. One is infeasible. The others produce the symbolic output presented in the figure. Input satisfying the condition under *Input* computes the output under *Output* if executed on the respective program version.

Technically, there are static [16] and dynamic [17, 18] approaches to symbolic execution. The former carry a symbolic state for each statement executed. The latter augment the symbolic state with a concrete state for the executed test input. A *symbolic state* expresses variable values in terms of the

		Input	Output
P		$i \leq 0$	$o = 0$
		$i > 0 \wedge j \leq 0$	$o = 1$
		$i > 0 \wedge j > 0$	$o = 2$
P'		$i \leq -1$	$o' = 0$
		$i > -1 \wedge j \leq -1$	$o' = 1$
		$i > -1 \wedge j > -1$	$o' = 2$

Fig. 4. Symbolic program summaries.

input variables and subsumes all feasible concrete values for the variable. A *concrete state* assigns concrete values to variables. System and library calls can be modeled as uninterpreted functions for which only dynamic SE can derive concrete output values for concrete input values by actually, concretely executing them [23].

In theory, path exploration can determine all feasible paths if it terminates. Yet, the number of paths grows exponentially to the number of conditional statements in the program. To attack this *path explosion problem*, it is possible to prune families of infeasible paths when one is encountered [24], group a set of feasible paths to a path family so as to explore only one member of a each group [15, 25, 26], massively parallelize the path exploration [27], and explore components of the program independently to compose the exploration results globally [28]. Further, scalable approaches are presented in combination with fuzz testing [29] and machine learning techniques [30].

3. CHANGE IMPACT ANALYSIS

Change impact analysis [31–34] can help to check whether and which program entities (including the output) are affected by syntactic program changes. The developer can focus testing efforts on affected program entities in order to more efficiently expose potential regression errors introduced by the changes.

Similar to dependence analysis, it is generally undecidable whether there exists input that exercises even a single changed statement [5] and not to mention that makes any behavioral difference observable. However, static analysis can approximate the potential existence of program paths that reach changes and propagate the semantic effects. Differential symbolic execution [35] allows a more precise analysis of the existence of program paths that can propagate the semantic effects of changes. Dynamic program analysis requires the existence of at least one such program path and can precisely determine the affected program entities and which changes are interacting.

3.1 Static Change Impact Analysis

Statically, we can determine (i) which statements are *definitely not affected* by a change [9, 10, 34], (ii) which statements are *probably affected* by a change [36], (iii) which set of changes do *definitely not semantically interfere* and can thus be tested in isolation [37, 38], and (iv) which statements remain, cease to, or begin to syntactically depend on a statement that is changed [39–41].

There are mainly two different syntactic approaches to statically compute the semantic difference introduced by the changes—text-based and dependency-based differencing. *Text-based differencing* [42–44] is a technique that given two program versions can expose changed code regions. This includes approaches that compare strings [43], as for instance the Unix utility `diff`, and approaches that compare trees [44]. Text-based differencing tools may efficiently identify textual differences but they cannot return information on code regions in the program that are affected by the changes.

Dependency-based differencing [39–41, 45] methods can compute the program entities affected by the changes. Using the static forward slice of the changed statements, we can compute those statements that are potentially affected by the change. Practically, this can be more than 90% of the statements in a program [33]. Still, every statement that is not in the static forward slice of any changed statement is definitely not affected by a change of that statement. Based on empirically justified assumptions, Santelices and Harrold [36] show how to derive the *probability* that the change of one statement has an impact on another given statement. Moreover, it is possible to check whether a set of changes potentially semantically interferes by computing the intersection of the static forward slices for each changed statement [37, 46]. If the static forward slices do not intersect, the set of changes can be tested in isolation.

For our running example, the static forward slices of the changes *ch*1 and *ch*2 in lines 2 and 6 are *not intersecting* at line 7 as shown in Fig. 5. In fact, only

Change Set	Interference Locations
{*ch*1, *ch*2}	6, 9, 10, 11
{*ch*1, *ch*3}	10, 11
{*ch*2, *ch*3}	10, 11
{*ch*1, *ch*2, *ch*3}	10, 11

Fig. 5. Potentially semantically interfering change sets.

*ch*1 may have a semantic effect on line 7. In contrast, the forward slices of both changed statements *are intersecting* at line 9, among others. Later in the text we show that *ch*1 and *ch*2 semantically interfere for input {0,0} because removing one change (by replacing the modified code with the original code for the change) alters the semantic effect of the other change on that execution. Therefore, both changes cannot be tested in isolation.

Using program slicing and reconstitution,[2] Horwitz [39] presents a technique to compute a program P_C for two program versions P and P' that exhibits all changed behaviors of P' w.r.t. P. The authors note that we cannot always assume to know the correspondence between the elements of the respective PDGs of both versions (P and P') and propose a solution using slice-isomorphism testing which executes in linear time [12]. The explicit (and automatic) tagging of every syntactic element is another solution to establish the correspondence of an element in the PDG in one version to an element in the PDG of another version [38]. Semantic differencing tools based on static dependency analysis were implemented by Jackson and Ladd [40] and more recently by Apiwattanapong et al. [45] and Loh and Kim [41]. However, while syntactic tools are efficient, they are often rather imprecise as the semantics of the programs are ignored. For instance, two syntactically very different pieces of code can always compute the same output for the same input. Yet, dependency-based tools will always report differences.

3.2 Dynamic Change Impact Analysis

Dynamically, given an input t, it is possible to determine (i) much more precisely which statements are affected by the (exercised) changes [31], (ii) whether and how the combined semantic effects of the exercised changes are propagated to the output [48, 49, 14]), and (iii) whether two subsets of the exercised changes are interacting [50].

Given the input t exercises only a single changed statement c, the *semantic effect* of c is propagated to another statement s (including the output) if s is not exercised in one but the other version or the values for the variables used in s are different in both versions (cf. [48, 51]). Two changes, c_1 and c_2, *interact* for the execution of t if removing one change (i.e., replacing the modified code with the original code for the change) alters the semantic effect of the other change on that execution. Santelices et al. [50] define and present a technique to compute change interaction. First, given two (sets of) changes

---

[2]A program is reconsituted when source code is generated from a dependence graph or program slice [47, 39].

c_1 and c_2, four program configurations are constructed—the modified program P', the modified program with c_1 being replaced by the original code $(P'\backslash c_1)$, the modified program with c_2 being replaced by the original code $(P'\backslash c_2)$, and the modified program with both changes being replaced by the original code $(P'\backslash\{c_1,c_2\})$. Second, the test case t is executed on all configurations to compute the execution traces $\pi(t,P')$, $\pi(t,P'\backslash c_1)$, $\pi(t,P'\backslash c_2)$, and $\pi(t,P'\backslash\{c_1,c_2\})\}$ augmented by variable values

$$\text{effect}(t,c_1,P') \leftarrow \text{diff}(\pi(t,P'),\pi(t,P'\backslash c_1)). \tag{1}$$

The semantic effect of c_1 on P' is computed as difference of the augmented execution traces when executing t on P' and on $P'\backslash c_1$

$$\text{interact}(t,c_1,c_2,P') \leftrightarrow ((\text{effect}(t,c_1,P') \neq \text{effect}(t,c_2,P'\backslash c_1))$$
$$\vee(\text{effect}(t,c_1,P'\backslash c_2) \neq \text{effect}(t,c_2,P'))). \tag{2}$$

Both changes c_1 and c_2 are interacting if the semantic effect of c_1 on P' is different from the semantic effect of c_2 on $P'\backslash c_1$ or the semantic effect of c_2 on P' is different from the semantic effect of c_1 on $P'\backslash c_2$.

An example of change interaction is depicted in Fig. 6. It shows two configurations—the modified program P' on the left-hand side and the modified program with $ch2$ being replaced by the original code, $P'\backslash ch2$, on the right-hand side. Input $t = \{0, 0\}$ exercises the changes $ch1$ and $ch2$ in lines 2 and 6 in both configurations. The *semantic impact* of $ch2$ on P' is the conditional statement in line 9 being evaluated in different directions in both configurations. As a result, input t produces output $o = 2$ in configuration P' and $o = 1$ in configuration $P'\backslash ch2$. The semantic impact of $ch1$ on $P'\backslash ch2$ is the conditional statement in line 5 being evaluated in different directions in both configurations. As a result, input t produces output $o = 1$

```
1   input(i=0,j=0);
2   a = i + 1;   //ch1
3   b = 0;
4   o = 0;
5   if(a > 0){
6       b = j + 1;//ch2
7       o = 1;
8   }
9   if(b > 0)    //(true)
10      o = o + 1;
11  output(o);
```

```
1   input(i=0,j=0);
2   a = i + 1;   //ch1
3   b = 0;
4   o = 0;
5   if(a > 0){
6       b = j;      //not ch2
7       o = 1;
8   }
9   if(b > 0)    //(false)
10      o = o + 1;
11  output(o);
```

Modified Version (P') P' without $ch2$ $(P'\backslash ch2)$

Fig. 6. Changes $ch1$ and $ch2$ interact for input $\{0, 0\}$.

in configuration $P'\backslash ch2$ and $o = 0$ in configuration $P'\backslash\{ch1, ch2\}$. Note, there does not exist any input for which $ch3$ has a semantic impact on any configuration. Both changes, $ch1$ and $ch2$ are *semantically interacting* for input $\{0, 0\}$ because the semantic impact of $ch2$ on P' is different from the semantic impact of $ch1$ on $P'\backslash ch2$ for t. Note, there does not exist any input for which $ch1$ or $ch2$ are interacting with $ch3$. Yet, in general it is undecidable whether there exists such an input that exercises a changed statement and propagates the semantic effects to another statement (incl. the output), or upon which two (sets of) changes are interacting.

3.3 Differential Symbolic Execution

Differential Symbolic Execution [35] can approximate those paths that potentially propagate the semantic effects of a change to the output. Exploiting the fact that the original and changed versions of a method are syntactically largely similar, the behavior of common code fragments is summarized as *uninterpreted functions*. In both versions the behavior of the changed method can be represented as *abstract program summaries*. An abstract summary consists of a set of partition-effect pairs. A partition-effect pair consists of a condition that is to be satisfied to observe the effect and an effect that computes the output in terms of the method input variables. Both, the condition and the output function can contain uninterpreted functions.

In our running example in Fig. 1 many code fragments are changed. Suppose that only the statement in line 10 is changed in the original program ($P'\backslash\{ch1, ch2\}$). Note, both versions P and $P'\backslash\{ch1, ch2\}$ are semantically equivalent (i.e., compute the same output for the same input). As depicted in Fig. 7, the behavior of the common code region from lines 2–8 is summarized as uninterpreted functions. In particular, the variable b used in line 9 is defined by the uninterpreted function $b(i, j)$ while o used in lines 11 and 12 is defined by the uninterpreted function $o(i, j)$.

To reveal the differential behavior of the changed version w.r.t. the original version, DSE allows to compute (partition-effects or functional) *deltas*

	Input	Output
P	$b(i, j) > 0$	$o = 2$
	$b(i, j) \leq 0$	$o = o(i, j)$
P'	$b(i, j) > 0$	$o' = o(i, j) + 1$
	$b(i, j) \leq 0$	$o' = o(i, j)$

Fig. 7. Abstract program summaries for P and P'\{ch1, ch2}.

upon both abstract summaries. For instance, if the conditions are the same but the effects are different in both versions and the computed delta does not contain an uninterpreted function, then every input satisfying the condition must expose a difference in program behavior. On the other hand, if the delta contains uninterpreted functions, then the behavior of the common code fragment has to be explored first. For instance, for the abstract summary in Fig. 7 DSE can show that if $b(i,j) > 0$ is satisfiable, the semantic effects of the changes may propagate to the output. However, in order to find an input that exposes a behavioral difference, first we have to check whether and for which values of i and j the condition $b(i,j) > 0$ can be satisfied. Second, we have to determine a value that satisfies $o \neq o'$ and thus $1 \neq o(i,j)$.

3.4 Change Granularity

When a new version of the program's source code is analyzed or tested, we may want to decompose this change from one version to the next into smaller "changes" which can be analyzed and tested in isolation. Syntactic change can be defined on different levels of granularity. For example, we can speak of changed components, features, classes, methods, code regions, statements, or of changed program dependencies.

In some cases changes cannot be tested in isolation and yield inconsistent program configurations. Zeller [52] distinguishes *integration failure*, for which one change requires another change that is not included in the configuration, *construction failure*, for which the configuration cannot be compiled, and *execution failure*, for which the program does not execute properly and the test outcome is unresolved.

Ren et al. [34] define change as cluster of changed statements that are required to avoid integration and construction failures. A program configuration can only contain every or no changed statement within a cluster of a selected change. In Fig. 8, change c_1 is adding method `inc` to a class. Change c_2 is adding a statement to that method. A configuration that contains c_2 must also contain c_1. The authors define several types of changes, such as adding, deleting, and changing methods or classes.

```
1  public class Test{
2    public int inc(int b){ //change c1: Add function
3      return b++; //change c2: Add statement
4    }
5  }
```

Fig. 8. Integration failure.

Jin et al. [53, 54] generate random test cases that are executed on both versions of a changed class. The authors note that the class interface should not change from one version to the next because the same unit test case cannot be executed on both versions simultaneously. Then, the test outcome is unresolved. Korel and Al-Yami [55] explain how to find the common input domain when the dimensionality of the input space changes.

Santelices et al. [50] define a code level change as "a change in the executable code of a program that alters the execution behavior of that program." The configuration $P'\backslash c$ is a syntactically correct version of P' where the original code of a change c replaces the modified code from that change.

4. REGRESSION TESTING

Regression testing is a technique that checks whether any errors are introduced when the program is changed. While static change impact analysis can reveal program elements that are not affected by the changes, regression testing should test those program elements which are potentially affected by the changes. In particular, software regression can only be observed for input that exposes a semantic difference in both programs.

Generally, regression testing is based on at least three assumptions: (i) the program behaves in a deterministic manner [17], (ii) the software tester is routinely able to check the correctness of the program output for any input [56], and (iii) an "adequate" regression test suite stresses much of the program's behavior, so that, when the program is changed and its behavior regresses, at least one test case fails upon execution on the changed program [2, 57].

4.1 Deterministic Program Behavior

A test case is meaningful only if executing the same input upon the same program always produces the same output—the program behavior is deterministic. Only then the output is representative for the test case and can be compared among program versions. Indeterminism can be introduced, for instance, by the program environment or concurrency.

The program environment can introduce indeterminism. Some authors [17] explicitly note that a library function, like an operating-system function or a function defined in the standard C library, is treated as unknown but deterministic black-box that cannot be analyzed but executed. In practice, this may not hold. Suppose that the analyzed program loads a file every time a test is executed. At one point the file is changed by a third party.

Suddenly the test fails even though the program did not change. An approach to model the execution environment is discussed by Qi et al. [58].

The behavior of concurrent programs can be considered indeterministic as well (cf. race conditions). This can be mitigated by constructing a finite model that considers all feasible schedules within which two or more threads can be executed concurrently and enumerate these schedules to determine for instance the existence of race conditions [59].

4.2 Oracle Assumption

In general, a software tester is not routinely able to check the correctness of the program output for any input. A mechanism that determines upon execution whether a test case passes or fails is known as *oracle*. In the context of evolving programs, an oracle further decides whether or not a behavioral difference exposed by a test case is intentional. If the difference is not intentional this test case would be a witness of regression. The oracle problem [56] postulates that an oracle that decides for every input whether the program computes the correct output is pragmatically unattainable and only approximate. Informally, the oracle problem denotes that even an expert may in some cases not be able to distinguish whether an observed functionality is a bug or a feature. However, there are types of errors that are generally acknowledged as such; for instance, exceptions, buffer overflows, array-out-of-bounds, or system crashes [17, 60, 54, 61]. These are called de facto or implicit oracles [56, 54]. Otherwise, it is possible to specify errors explicitly as assertion-, property-, or specification violations [62–64]. In some cases, the same functionality is implemented more than once to compare the output [65] or the program is run on "simplified" input data to accurately assess the "simple" output [56].

The oracle problem affects specifically automated test generation, debugging, and bugfixing techniques. For instance, an automated bugfixing technique can correct the (buggy) program only relative to explicitly specified or known errors. In a recent work, Staats et al. [66] point out that empirical software testing research should explicitly consider the definition of oracles when presenting the empirical data in order to better evaluate the efficacy of a testing approach and allow for comparison by subsequent studies.

4.3 Code Coverage as Approximation of Adequacy

The measure of code coverage *approximates* the adequacy of a test suite to cover much of the program behavior [57]. A test suite is 100% *code*

coverage-adequate w.r.t. a coverage criterion if all instances of the criterion
are exercised in a program by at least one test case in the test suite [67]. A
statement coverage-adequate test suite requires that every statement in the
program is exercised by at least one test case in the test suite. Decision coverage requires that the condition in every control structure is evaluated both,
to true and false. A path coverage-adequate test suite exercises every feasible
path from program entry to exit at least once [67]. The measure of code
coverage (excepting path coverage) can often be absolutely computed using
syntactic representations of the source code, such as the nodes and edges in a
PDG. For instance, a test suite is 50% statement coverage-adequate, if all test
cases in the test suite exercise exactly half of the statements in the program.
For our running example, the test suite T_{RE} in Eqn 3 covers every path in
both program versions (cf. Fig. 4)

$$
T_{RE} = \left\{ \begin{array}{rrr} \{ & -2, & -2 \ \}, \\ \{ & 2, & -2 \ \}, \\ \{ & 2, & 2 \ \} \end{array} \right\}. \tag{3}
$$

Generally, it is undecidable whether there exists a 100% coverage-
adequate test suite for a given program and a given coverage criterion because
it is undecidable whether there exists an input that exercises a path containing
a given syntactic program artifact [5]. While code-coverage can often be efficiently computed for a test suite w.r.t. a finite amount of syntactic program
artifacts, there are other measures to assess the test suite adequacy, such as
fault-based [68, 69, 22], change-based [70], or "behavioral" [57] criteria. The
efficacy of the different measures can vary and has been compared [71–74].

The approximation of the amount of covered behavior by the amount
of covered code may not properly quantify the capability of a test suite to
reveal regression errors. Specifically, a code coverage-adequate test suite may
not inspire confidence in the correctness of the program [75] and may not
perform significantly better than random generated test cases in terms of
revealing program errors [74–77]. Weyuker et al. [74] observe that while
a test case represents one or more equivalence classes in the input space
of a program,[3] such an equivalence class may not be homogeneous w.r.t.
failure—if one test case fails, every input in the same class fails. For instance,
it is not true that if a test case exercises some branch (which it may represent)
and exposes an error, then every input exercising the same branch exposes an
error. This may be sufficient motivation to study semantic coverage criteria

---

[3]E.g., for branch coverage, one class may represent every input exercising a certain branch.

that can be used to better assess the adequacy of regression test suites (in terms of the covered input space) [78]. As for our running example, the regression test suite T_{RE} in Eqn 3 exercises every path in both versions. However, it does not expose any behavioral difference when comparing the output upon execution in both versions. As software regression is observable only for input that exposes a behavioral difference, we can conclude that even a path coverage-adequate test suite may not expose software regression.

5. REDUCTION OF REGRESSION TEST SUITES

In order to gain confidence that program changes did not introduce any errors, regression test suites are executed recurringly. The number of test cases can greatly influence the execution time of a test suite. When the program is changed, we can choose to execute only relevant test cases that actually execute the changed code regions and are more likely to expose regression errors. Similarly, we can permanently remove test cases that are irrelevant w.r.t. some measure of test suite adequacy.

5.1 Selecting Relevant Test Cases

Given a test suite, when the program is changed, only those test cases may be selected that actually stress the changed functionality and can expose software regression [79, 80, 34, 81]. On the other hand, test cases that do not exercise the program changes cannot expose software regression that is introduced by these changes. Ideally, executing only the selected test cases reduces the testing time while preserving the capability to reveal regression errors.

For example, Ren et al. [34] present a tool that given a test suite can determine test cases that do judicably not exercise any changed statement. For the analyzed subjects, on average 52% of the test cases were *potentially affected* by the changes; each test case by about 4% of the changes. Furthermore, given a test suite, the tool can ascertain which changed statements are judicably not executed by any test case. The test suite should be augmented by test cases that exercise these statements to decide whether this change introduced any regression errors.

Graves et al. [79] empirically compare several test selection techniques. The *minimization technique* chooses only those test cases that cover the modified or affected parts of the program. It produces the smallest and least effective test suite. The *safe technique* selects all test cases in the original test suite that can reveal faults in the program. This technique was shown to find all faults while selecting 60% of the test cases on the median. The *ad-hoc or random technique* selects test cases on a (semi-) random basis. The random technique produced slightly larger test suites than the minimization technique but on average yielded fault detection results equivalent to those of the minimization technique with little analysis costs. Furthermore, randomly selected test suites could be slightly larger than a safely selected test suite but nearly as effective.

5.2 Removing Irrelevant Test Cases

Test cases in a large test suite that are redundant in some respect may be removed completely [82–84]. Ideally, test suite reduction decreases the execution time of recurring regression testing while preserving the capability to reveal regression errors. Considering test cases as representatives of equivalence classes, it is possible to remove those test cases that represent the same equivalence class without reducing the current measure of adequacy. For instance, given a 95% branch coverage-adequate test suite T, test cases are removed from T until the removal of one more test case also reduces the branch-coverage of T to less than 95%. Based on their empirical results, Rothermel et al. [85] conclude that "test suite minimization can provide significant savings in test suite size. These savings can increase as the size of the original test suites increases, and these savings are relatively highly correlated (logarithmically) with test suite size."

However, the reduction of a test suite w.r.t. a code coverage criterion has a negative impact on the capability of a test suite to reveal a fault [86, 87]. Hao et al. [88] observe that the reduction w.r.t. statement coverage incurs a loss in fault-detection capability from 0.157 to 0.592 (with standard deviations from 0.128 to 0.333) for the analyzed subjects. In other words, about 16–60% of the faults originally detected become unexposed using the reduced test suite. Yu et al. [86] empirically determine that the reduction of a test suite w.r.t. statement coverage increases the fault localization expense by about 5% on average for the analyzed subjects. In other words, given original test suite T and the test suite T' that is reduced w.r.t. statement coverage, if Tarantula[4] were to pinpoint a single statement as probable fault location using T, then Tarantula would require the tester to examine 5% of the source code as probable fault location using T'. In a recent work, Hao et al. [88] propose a test suite reduction technique that removes test cases from the test suite while maintaining the capability to reveal faults above a user-defined threshold.

---

[4]Tarantula is an automatic fault-localization technique [89].

6. AUGMENTATION OF REGRESSION TEST SUITES

An existing test suite should be augmented by relevant test cases (i) to better satisfy a given test suite adequacy criterion, and (ii) to expose behavioral differences introduced by changes to the program. Only such difference-revealing test cases can potentially expose software regression.

There are automatic test generation techniques to better satisfy coverage-based [90–92, 29], fault-based [93, 94, 22], and "behavioral" [57] adequacy criteria. Approaches to generate test cases that expose a behavioral difference in two program versions can be coarsely distinguished into three classes. *Syntactic approaches* [51, 95, 96] aim to generate input that first reaches at least one change, then infects the program state, and thereupon propagates its semantic effect to the output. *Semantic approaches* [35, 15, 110] use a form of program summaries to find input that exposes a difference. *Random approaches* [54, 97] randomly generate test cases that may or may not expose a difference when executed on both versions.

6.1 Reaching the Change

Search-based test generation techniques [19, 98] aim to generate test cases that reach specified targets in the program. These targets can be coverage goals to increase code-coverage [90, 91, 29], program changes [95, 51, 21, 22, 94], or specified program faults like assertions [62, 55], exceptions [99, 60], and (functional) properties [63, 64]. Korel and Al-Yami [55] present a technique that given two program versions reduces the problem of generating input that exposes a behavioral difference to the problem of reaching an assertion.

It is generally undecidable whether there exists an input that reaches a change [5]. Practically, we can generate test cases to search for such input. If we can assign a given input some measure of *distance* to the change, then we can apply search strategies that reduce this distance. The distance of a test case t to a changed statement c can be defined, for instance, based on the length of the control dependency chain from c to those branches exercised by t that are not evaluated in favor of the execution of c, that is, have to be negated in order to reach c.

Local search strategies, such as hill climbing [62, 100], monotonically reduce this distance. Random restart procedures [22] can prevent the search strategy to get stuck in a local minimum distance. Ferguson and Korel [100] introduce the Chaining Approach (CA) that leverages data- and control dependencies to generate input that reaches a target by identifying and exercising a necessary sequence of nodes beforehand. Given a target c, CA analyzes the program dependency graph to find program input that exercises c. The target c can be reached only if those nodes upon which c control-depends are evaluated in favor of the execution of c. Given a node p upon which c control-depends is not evaluated in favor of c for some input t, then CA will generate input for which p is negated. If p cannot be negated by input exercising the same path (i.e., the same sequence than t of nodes in the CFG), then p is marked as the *problem node*. "The chaining approach finds a set $LD(p)$ of last definitions of all variables used at problem node p. By requiring that these nodes are executed prior to the execution of problem node b, the chances of altering the flow execution at problem node p may be increased" [100]. Effectively, the nodes in $LD(p)$ become intermediate target nodes. This sequence of (intermediate) target nodes is called *event sequence* (or chain).

We explain the chaining approach for our running example in Fig. 9. Suppose, we want to generate an input for the modified version P' that exercises the changed statement in line 10. The CA may start with random input $\{-2, -2\}$ as shown on the left-hand side. CA determines the branch in line 9 as problem node. The only variable used in the condition is b which is defined in lines 3 and 6. So, CA designates the statement in line 6 as the intermediate target which is guarded by the branch in line 5. This branch is evaluated to *false*. To negate this branch, CA has to compute an input so that $i + 1 > 0$ (using function minimization). Thus, the next input may be $\{2, -2\}$ as shown on the right-hand side. The branch in line 9 guarding the target in line 10 can be negated by input exercising the same path than $\{2, -2\}$. In particular, CA computes an input so that $i + 1 > 0 \wedge j + 1 > 0$ which is satisfied by test case $\{2, 2\}$.

```
1   input(i=-2,j=-2);
2   a = i + 1;
3   b = 0;
4   o = 0;
5   if(a > 0){   //(4)(false)
6      b = j + 1;//(3)intermed.
7      o = 1;
8   }
9   if(b > 0)      //(2)problem node
10     o = o + 1;//(1)target
11  output(o);
```

```
1   input(i=2,j=-2);
2   a = i + 1;
3   b = 0;
4   o = 0;
5   if(a > 0){   //(true)
6      b = j + 1;
7      o = 1;
8   }
9   if(b > 0)      //(2)(false)
10     o = o + 1;//(1)target
11  output(o);
```

Search state with input $\{-2, -2\}$ Search state with input $\{2, -2\}$

Fig. 9. Chaining approach explained for modified program P'.

Search strategies based on genetic algorithms [101], choose the "fittest" set of inputs from one generation as "seed" for the next generation to find a global minimum distance. Search strategies based on counterexample-guided abstraction refinement [63, 102, 64] try to prove that no such input exists in an abstract theory. If instead a (possibly spurious) counterexample is found, it continues to prove the absence of a counterexample in a refined theory. This repeats until either its absence is proven or a concrete (non-spurious) counterexample is found. A particular kind of search strategies seeks to cover a *set of targets* at once or in a given sequence [91, 20, 22].

To optimize the search, it is possible to reduce the search-space in a sound [98, 26, 24, 96] and approximative manner [103, 25], search distinct program components independently and compose the results [28, 64], or execute the search strategy on multiple instances in parallel [27]. Yet, since the problem is undecidable in general, the search for an input that reaches a change may never terminate in some cases [63].

Another practical approach to find input that reaches a change is the random generation of program input [104, 77, 105, 54]. Arcuri et al. [106] analytically determine that the time to reach all of k targets by random test generation is lower-bounded by the linearithmetic function, $O(k * log(k))$.

6.2 Incremental Test Generation

Given only the changed statements in the changed program P', incremental test generation is concerned with testing the code regions that are affected by the changes. On the one hand, test cases that do not exercise a changed statement cannot reveal a behavioral difference [34]. On the other hand, test cases that do exercise one or more changed statements may or may not yield an observable behavioral difference [51, 107, 21]. In fact, one study [22] finds that only 30–53% of the test cases that do exercise a changed statement are difference-revealing for the analyzed whole programs. In our running example every input exercises at least one change (*ch*1) but only about 2–25% exposes a semantic difference (cf. Fig. 17).

In general, every statement in the static forward slice of a changed statement is potentially affected by the change [10]. Hence, one can direct the path exploration of P' explicitly toward the changed statements in order to exercise program paths that are affected by the changes and increase the likelihood to observe a behavioral difference [21]. Vice versa, one can avoid the exploration of paths in P' that will not stress a changed code region and are unlikely to propagate the semantic effect of a change [96].

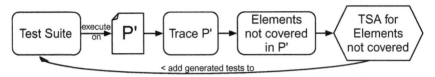

Fig. 10. Re-establishing code coverage.

Upon program change, the code coverage of an existing test suite may decrease. As outlined in Fig. 10, Xu et al. [90, 92] firstly apply a test selection technique to find all test cases that are affected by the changes. Secondly, these test cases are executed on the changed program to determine syntactic program artifacts that are not covered (anymore). Lastly, the authors seek to re-establish the code-coverage by generating test cases that exercise those syntactic program elements that are not covered in P' reusing the selected test cases.

The analysis of only a single version, either P or P', is insufficient to expose all behavioral differences. Even input exercising the same affected path in P' may exercise multiple, different paths in the original version P [108]. As a result, the semantic interaction [50] of a set of changes may or may not be observed at the output, even if every affected path is exercised. As for our running example, the test suite T_{RE} in Eqn 3 exercises every path in both program versions. However, this test suite does not expose any behavioral difference when comparing the output upon execution in both versions.

6.3 Propagating a Single Change

One may ask: What is the semantic impact of a change onto the program? Does it introduce a bug? Since it is undecidable whether there exists input that exercises the changed statement [5], it is also undecidable whether there exists an input that reveals a behavioral difference and not to mention software regression. However, given both program versions P and P' we can search for input that (1) reaches the changed statement, (2) infects the program state, and (3) propagates the semantic effect to the output [95, 48, 93].

Santelices et al. [109, 51] describe a technique that derives requirements for new test cases to propagate the semantic effect of the exercised change to a user-specified minimum distance (in terms of static dependence chains starting at the changed statement). The tester can use these requirements to write a test case that is more likely to reveal different behavior in the changed version than a test case that merely executes the change. Using text-based differencing, the algorithm finds the changed statement in the original program

P and modified version P'. Then, by means of (partial, dynamic) symbolic execution the path condition and symbolic state for those statements following the changed statement are computed. The path conditions and symbolic states of the corresponding statements are compared for P and P' and testing requirements are derived.

Qi et al. [95] generate a test case t, so that t executes a given change c and the effect of c is observable in the output produced by t. The test case t can be considered a *witness* of the behavioral difference introduced by c in both program versions. The underlying algorithm works as follows.

First, using an efficient hill-climbing search strategy, input that reaches the changed statement is generated. For optimization, all test cases in an existing test suite are executed and respective path conditions are derived. A distance function determines the probability of an input to reach a change and imposes an order over the test inputs. Always taking the input "closest" to the change, the respective path condition is manipulated to generate new input t_{new} that minimizes the distance to the changed statement for the execution of t_{new} on P'. This repeats until the distance is zero and the change is reached.

Second, using the Change Effect Propagation Tree (CEPT), the semantic effect of the changed statement is propagated to the output. The semantic effect of a change is observable for an input t in a variable v along the path (and ultimately at the output) if v has a different value for the execution of t on P than of t on P' (i.e., execute$(t, P', v) \neq$ execute$(t, P' \backslash c, v)$). The CEPT identifies terminating locations of effect propagation for each execution. The CEPT represents why a change cannot be propagated any further. For example, a variable v that carries the semantic effect of the change is redefined without using v anywhere else before. The authors determine three different reasons for propagation termination and handle them accordingly. The path condition is modified to drive the execution along a path that ensures propagation if possible. This repeats until the semantic effect of c is observable in the output.

A simplification of the process is shown in Fig. 11. For each diagram the left (and black) line depicts the possible augmented execution trace[5] for the original program P. The right (and red) line shows the augmented history for the modified program P'. A deviation of both lines indicates that the same input begins to produce different states in both versions at this point.

---

[5]The augmented execution trace is the sequence of executed program statements plus respective, relevant program states [50].

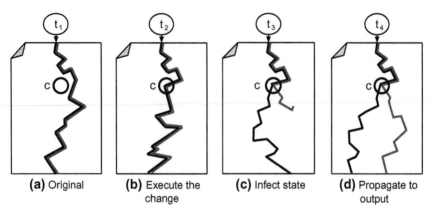

(a) Original   **(b)** Execute the change   **(c)** Infect state   **(d)** Propagate to output

Fig. 11. Generating input that satisfies the PIE principle.

Test	Input	P	P'
t_1	$\{0,-1\}$	$o = 0$	$o' = 1$
t_2	$\{0,\ \ 0\}$	$o = 0$	$o' = 2$

Fig. 12. Behavioral differences between P and $P' \setminus \{ch1, ch2\}$.

The generated concrete test case is only one witness of the changed behavior. The syntactic change could modify the behavior of the program in more than one way. Qi et al. give an approach that shows *some path* that exposes changed behavior due to the change. A regression error may only be exposed on another path leading to the output. Figure 12 shows two test cases witnessing a behavioral difference between original version P and a configuration for which only the single change $ch1$ is applied to P. In theory, even for a single change there may be infinitely many paths that exercise the changed statement and produce a different output in both versions (e.g., if a loop condition depends on the input).

6.4 Propagation of Multiple Changes

When multiple statements are changed, they may semantically interfere [38] or interact [50] subtly and unintendedly when executed on some input but not on others. Program changes potentially semantically interfere if the static forward slices of the changed statements intersect in the changed program [37, 38]. Then, the changes cannot be tested in isolation. For a given input, the semantic effect of one change onto a statement may be masked or augmented by the semantic effect of another change onto that statement.

Santelices et al. [51] discuss the feasibility of the approach of propagating a single change to a minimal distance [109] in the presence of multiple (non-interfering) changes. For each change there has to exist a path from the program entry to the changed statement that does not contain another changed statement. This ensures that the semantic effect of an "earlier" change is not accidently propagated to a statement that is supposed to be infected by the given change.

Harman et al. [22] call potentially semantically interfering sets of code changes "higher-order mutants" and aim to generate a test case that for a given higher-order mutant produces different output in both versions. First, using the control dependence graph of the changed program, the technique computes a path that may execute all changes. Then, using this information and a hill climbing algorithm with random restart, the technique generates a test case that exercises every changed statement of the given higher-order mutant. Lastly, the technique searches paths that are more likely to propagate the combined semantic effects to the output.

Given a set of changes C, there are $2^{|C|-1}$ subsets of C that potentially semantically interfere and have to be tested. For example, our running example has three changes yielding four possibly interfering change sets (cf. Fig. 5). Yet, even for a single subset, the search for a difference-revealing test case may never terminate, which renders this procedure prohibitively expensive. Even if the search yields an input that produces different output on both program versions, this input may not be a witness of software regression.

6.5 Semantic Approaches to Change Propagation

While syntactic techniques seek to explicitly reach at least one change and propagate its semantic effect to the output, semantic techniques compute differences based on the transformation functions of original and modified program version [35, 78, 110]. Path exploration based on the symbolic output can reveal the transformation function of a program [15]—the *symbolic program summary*. This summary is an (incomplete) list of input partitions. Each input in the same partition computes the same symbolic output. Given the program summaries of two program versions, a behavioral difference is exposed by input that computes different output. In other words, if for overlapping input partitions the output is computed differently, then every input in this intersection exposes a behavioral difference.

Figure 13 lists the symbolic program differences for the two versions in our running example. The respective symbolic summaries are shown in Fig. 4. The intersection is found by conjoining every input condition and

$\{i, \ j\}$	Input	Output	Diff
$\{-1, \ 0\}$	$i \leq -1$	$o = o' = 0$	
$\{0,-1\}$	$i > -1 \wedge i \leq 0 \wedge j \leq -1$	$o = 0 \wedge o' = 1$	x
$\{0, \ 0\}$	$i > -1 \wedge i \leq 0 \wedge j > -1$	$o = 0 \wedge o' = 2$	x
$\{1,-1\}$	$i > 0 \wedge j \leq -1$	$o = o' = 1$	
$\{1, \ 0\}$	$i > 0 \wedge j > -1 \wedge j \leq 0$	$o = 1 \wedge o' = 2$	x
$\{1, \ 1\}$	$i > 0 \wedge j > 0$	$o = o' = 2$	

Fig. 13. Symbolic program difference for P and P'.

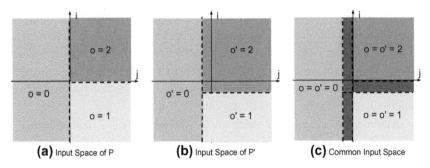

(a) Input Space of P   **(b)** Input Space of P'   **(c)** Common Input Space

Fig. 14. Visualization of overlapping input space partitions.

testing for satisfiability. Note that input exercising the same path in $P, i \leq 0$, can exercise three paths in P'.

The visualization of the input space partitioning is shown in Fig. 14. The set of all values for input variables i and j forms a two-dimensional vector space. Figure 14a shows the input space of the original program P partitioned in terms of the output values. Figure 14b depicts the input space of the modified program P' partitioned in terms of the output values. The overlapping of the partitioning of both input spaces is visualized in Fig. 14c for the common input space. The plane, shown in dim gray, represents input that executed on both versions compute different output.

By summarizing the behavior of code blocks that are common in both versions as uninterpreted functions, we can derive the *abstract program summary* for each program [35]. The abstract summaries can be used to compute the partition-effect deltas. Such a delta reflects input partitions and their associated effects, present in one version that are not present in the other version of a program. Figure 15 shows the functional delta computed for the abstract summaries in Fig. 7. If there exists input that renders $b(i,j) > 0$ satisfiable, then there may be a difference in output observable.

	Input	Output
$\Delta_{P,P'\setminus\{ch1,ch2\}}$ $= \Delta_P$	$b(i,j) > 0$	$o = 2$
$\Delta_{P'\setminus\{ch1,ch2\},P}$ $= \Delta_{P'\setminus\{ch1,ch2\}}$	$b(i,j) > 0$	$o' = o(i,j) + 1$

Fig. 15. Partition-effect deltas for P w.r.t. $P'\setminus$ {ch1,ch2}, and vice versa.

More specifically, if there exists an assignment to i and j that renders $b(i,j) > 0 \wedge o(i,j) \neq 1$ satisfiable, then this input is a witness of semantic difference. Note, for the versions P and $P'\setminus\{ch1,ch2\}$ there does not exist such an input.

Korel and Al-Yami [55] present a technique that given two program versions reduces the problem of generating input that exposes a behavioral difference to the problem of reaching an assertion. The technique generates a test driver that wraps both program versions and adds the assertion that both versions compute the same output values. Then the technique searches for a witness that violates that assertion using a hill climbing strategy similar to the one presented in Ref. [62]. This witness is a difference-revealing input for both program versions.

6.6 Random Approaches to Change Propagation

Random test generation techniques can provide test cases that, when executed on both program versions, reveal a difference [54, 97]. The procedure is depicted in Fig. 16.

Jin et al. [54] present a technique to generate random input, execute it on both versions, and report such cases that yield different output. The proposed technique determines the syntactic difference between two versions through static analysis. Leveraging Randoop [105] as random test generation engine, a large body of test inputs are generated for the set of changed classes. The generated test suite is then run on both versions of those classes, the output compared, and the differences in output reported as behavioral

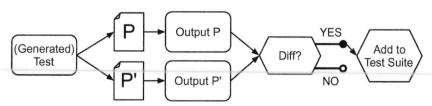

Fig. 16. Behavioral regression testing.

Input	Diff	Solutions	Probability
$i \leq -1$		$(2^{31} - 1)*(2^{32} - 1)$	~ 0.5
$i > -1 \wedge i \leq 0 \wedge j \leq -1$	x	$1*(2^{31} - 1)$	2^{-33}
$i > -1 \wedge i \leq 0 \wedge j > -1$	x	$1*2^{31}$	2^{-33}
$i > 0 \wedge j \leq -1$		$(2^{31} - 1)*(2^{31} - 1)$	~ 0.25
$i > 0 \wedge j > -1 \wedge j \leq 0$	x	$(2^{31} - 1)*1$	2^{-33}
$i > 0 \wedge j > 0$		$(2^{31} - 1)*(2^{31} - 1)$	~ 0.25

Fig. 17. Random input reveals a difference with probability $3 * 2^{-33}$.

differences. A challenge of the technique is the change of method signatures from one version to the next. The same unit test cases cannot be executed on both versions. While this is a scalable approach, Santelices and Harrold [107] empirically show that regression errors in low-probability domains are likely to stay undetected when sampling a normal distribution.

For our running example, Fig. 17 shows the probability to reveal a difference if we consider i and j to be 32-bit signed integers that are randomly generated. The probability to randomly generate difference-revealing test cases is $3 * 2^{-33}$ (about two magnitudes smaller than winning the UK-lottery). In practical terms, setting a bound to -2 and 2 (or -10 and 10), the probability to generate a difference-revealing test case would be 0.28 (or 0.07, respectively) even though every input exercises at least one change.

7. CONCLUSION

Software testing remains the most important form of software validation despite advances in program analysis, model checking, and theorem proving via Satisfiability Modulo Theory (SMT) solving. Each of these techniques provide a different perspective of program checking. Program analysis tries to find "bugs" by inferring program properties. Model checking attempts to find problematic test inputs by searching a large search space. Finally, theorem proving is inherently different—it attempts to prove programs correct via deduction, rather than generating potentially problematic test cases.

Despite the huge advances in constraint solving, search space representation and exploration (for model checking) and theorem proving—testing still remains hugely popular. Why? This is because of the concrete outcome from testing as an activity—once the test cases are generated—the activity of testing immediately points us to a potential bug if the test case fails.

This aspect of testing is further magnified, when we want to validate a new program version against absence of regressions. Notably, while testing a single program version—a notion of "expected output" is needed to validate the observed program output. However, while testing a program version against a previous version to check for regressions—we can often compare the output of the current program version with the previous program version. Thus, testing a program version against regressions from the previous version can immediately lead us to failing tests which expose potential regressions.

We have shown in this article that technologies useful for program analysis and theorem proving—namely program dependence analysis and SMT constraint solving can be gainfully employed in test generation for regression testing. In particular, the combination of program dependency analysis, symbolic execution, and SMT constraint solving can lead to powerful test generation methods as discussed in this article. Indeed, such combinations can be useful for fast efficient test generation, as well as test generation to expose changes across program versions. In particular, one can explore outgoing program dependencies from the changes (across two program versions) and perform symbolic execution along these dependencies to calculate a formula. This formula is then solved using a SMT solver to yield a test input which exercises the outgoing dependencies from a program change, and propagates the effect of the change to the program output.

In terms of future perspectives, we feel that analysis methods combining dependency analysis and symbolic execution hold promise in terms generating test cases that can give stronger semantic guarantees for a greater set of inputs. As shown in the recent work [15], one can use a program dependency chain to group program paths—paths which share the same dependency chain are put in the same partition. Symbolic execution along the dependency chains then leads to a formula representing the set of inputs whose execution traces share the same dynamic dependency chain. By solving such formulae, we can generate representative test inputs which exercise different dependency chains. For example, a test case exercising a relevant slice computed w.r.t. the output can represent every input exercising the same relevant slice and thus computing the same symbolic output [15]. Furthermore, for multiple program versions, in order to detect regression, we can inspect the change of the dynamic dependencies by observing the change of path and input partitioning, respectively [110]. This amounts to a new style of testing for program versions, where we not only detect inputs observing regression errors, but also groups of inputs which observe the same regression error.

REFERENCES

[1] R.C. Seacord, D. Plakosh, G.A Lewis, Modernizing Legacy Systems: Software Technologies, Engineering Process and Business Practices, Addison-Wesley Longman Publishing Co., Inc., Boston, MA, USA, 2003.

[2] G.J. Myers, C. Sandler, The Art of Software Testing, John Wiley & Sons, 2004.

[3] Z. Gu, E.T. Barr, D.J. Hamilton, Z. Su, Has the bug really been fixed? in: Proceedings of the 32nd ACM/IEEE International Conference on Software Engineering, ICSE '10, vol. 1, ACM, New York, NY, USA, 2010, pp. 55–64.

[4] Z. Yin, D. Yuan, Y. Zhou, S. Pasupathy, L. Bairavasundaram, How do fixes become bugs? in: Proceedings of the 19th ACM SIGSOFT Symposium and the 13th European Conference on Foundations of Software Engineering, ESEC/FSE '11, ACM, New York, NY, USA, 2011, pp. 26–36.

[5] A. Goldberg, T.C. Wang, D. Zimmerman, Applications of feasible path analysis to program testing, in: Proceedings of the 1994 ACM SIGSOFT International Symposium on Software Testing and Analysis, ISSTA '94, ACM, New York, NY, USA, 1994, pp. 80–94.

[6] M.C. Thompson, D.J. Richardson, L.A Clarke, An information flow model of fault detection, in: Proceedings of the 1993 ACM SIGSOFT International Symposium on Software Testing and Analysis, ISSTA '93, ACM, New York, NY, USA, 1993, pp. 182–192.

[7] A. Podgurski, L.A Clarke, A formal model of program dependences and its implications for software testing, debugging, and maintenance, IEEE Trans. Softw. Eng. 16 (1990) 965–979.

[8] K.J. Ottenstein, L.M. Ottenstein, The program dependence graph in a software development environment, in: Proceedings of the First ACM SIGSOFT/SIGPLAN Software Engineering Symposium on Practical Software Development Environments, SDE 1, ACM, New York, NY, USA, 1984, pp. 177–184.

[9] M. Weiser, Program slicing, in: Proceedings of the 5th International Conference on Software Engineering, ICSE '81, 1981, pp. 439–449.

[10] S. Horwitz, T. Reps, D. Binkley, Interprocedural slicing using dependence graphs, ACM Trans. Program. Lang. Syst. 12 (1990) 26–60.

[11] B. Korel, J. Laski, Dynamic program slicing, Inf. Process. Lett. 29 (1988) 155–163.

[12] S. Horwitz, T. Reps, Efficient comparison of program slices, Acta Inf. 28 (9) (1991) 713–732.

[13] G.A. Venkatesh, The semantic approach to program slicing, in: Proceedings of the ACM SIGPLAN 1991 Conference on Programming Language Design and Implementation, PLDI '91, ACM, New York, NY, USA, 1991, pp. 107–119.

[14] T. Gyimóthy, A. Beszédes, I. Forgács, An efficient relevant slicing method for debugging, in: Proceedings of the 7th European Software Engineering Conference Held Jointly with the 7th ACM SIGSOFT International Symposium on Foundations of Software Engineering, ESEC/FSE-7, Springer-Verlag, London, UK, 1999, pp. 303–321.

[15] D. Qi, H.D. Nguyen, A. Roychoudhury, Path exploration based on symbolic output, in: Proceedings of the 19th ACM SIGSOFT Symposium and the 13th European Conference on Foundations of Software Engineering, ESEC/FSE '11, ACM, New York, NY, USA, 2011, pp. 278–288.

[16] J.C. King, Symbolic execution and program testing, Commun. ACM 19 (1976) 385–394.

[17] P. Godefroid, N. Klarlund, K. Sen, Dart: directed automated random testing, SIGPLAN Not. 40 (2005) 213–223.

[18] K. Sen, D. Marinov, G. Agha, Cute: a concolic unit testing engine for c, SIGSOFT Softw. Eng. Notes 30 (2005) 263–272.

[19] P. McMinn, Search-based software test data generation: a survey: research articles, Softw. Test. Verif. Reliab. 14 (2) (2004) 105–156.

[20] W. Jin, A. Orso, Bugredux: reproducing field failures for in-house debugging, in: Proceedings of the 2012 International Conference on Software Engineering, ICSE 2012, Piscataway, NJ, USA, 2012, pp. 474–484.

[21] S. Person, G. Yang, N. Rungta, S. Khurshid, Directed incremental symbolic execution, in: Proceedings of the 32nd ACM SIGPLAN Conference on Programming Language Design and Implementation, PLDI '11, 2011, pp. 504–515.

[22] M. Harman, Y. Jia, W.B. Langdon, Strong higher order mutation-based test data generation, in: Proceedings of the 19th ACM SIGSOFT Symposium and the 13th European Conference on Foundations of Software Engineering, ESEC/FSE '11, ACM, New York, NY, USA, 2011, pp. 212–222.

[23] P. Godefroid, Higher-order test generation, in: Proceedings of the 32nd ACM SIGPLAN Conference on Programming Language Design and Implementation, PLDI '11, ACM, New York, NY, USA, 2011, pp. 258–269.

[24] M. Delahaye, B. Botella, A. Gotlieb, Explanation-based generalization of infeasible path, in: Proceedings of the 2010 Third International Conference on Software Testing, Verification and Validation, ICST '10, IEEE Computer Society, Washington, DC, USA, 2010, pp. 215–224.

[25] R. Santelices, M.J. Harrold, Exploiting program dependencies for scalable multiple-path symbolic execution, in: Proceedings of the 19th International Symposium on Software Testing and Analysis, ISSTA '10, ACM, New York, NY, USA, 2010, pp. 195–206.

[26] P. Boonstoppel, C. Cadar, D. Engler, Rwset: attacking path explosion in constraint-based test generation, in: Proceedings of the Theory and Practice of Software, 14th International Conference on Tools and Algorithms for the Construction and Analysis of Systems, TACAS'08/ETAPS'08, Springer-Verlag, Berlin, Heidelberg, 2008, pp. 351–366.

[27] M. Staats, C. Păsăreanu, Parallel symbolic execution for structural test generation, in: Proceedings of the 19th International Symposium on Software Testing and Analysis, ISSTA '10, ACM, New York, NY, USA, 2010, pp. 183–194.

[28] S. Anand, P. Godefroid, N. Tillmann, Demand-driven compositional symbolic execution, in: Proceedings of the Theory and Practice of Software, 14th International Conference on Tools and Algorithms for the Construction and Analysis of Systems, TACAS'08/ETAPS'08, Springer-Verlag, Berlin, Heidelberg, 2008, pp. 367–381.

[29] P. Godefroid, M.Y. Levin, D.A. Molnar, Automated whitebox fuzz testing, in: Proceedings of the Network and Distributed System Security Symposium, NDSS '08, The Internet Society, 2008.

[30] M. Davies, C. Pasareanu, V. Raman, Symbolic execution enhanced system testing, in: R. Joshi, P. Müller, A. Podelski (Eds.), Verified Software: Theories, Tools, Experiments, Lecture Notes in Computer Science, vol. 7152, Springer, Berlin/Heidelberg, 2012, pp. 294–309.

[31] S. Lehnert, A taxonomy for software change impact analysis, in: Proceedings of the 12th International Workshop on Principles of Software Evolution and the 7th Annual ERCIM Workshop on Software Evolution, IWPSE-EVOL '11, ACM, New York, NY, USA, 2011, pp. 41–50.

[32] T. Apiwattanapong, A. Orso, M.J. Harrold, Efficient and precise dynamic impact analysis using execute-after sequences, in: Proceedings of the 27th International Conference on Software Engineering, ICSE '05, ACM, New York, NY, USA, 2005, pp. 432–441.

[33] A. Orso, T. Apiwattanapong, M.J. Harrold, Leveraging field data for impact analysis and regression testing, SIGSOFT Softw. Eng. Notes 28 (5) (2003) 128–137.

[34] X. Ren, F. Shah, F. Tip, B.G. Ryder, O. Chesley, Chianti: a tool for change impact analysis of java programs, in: Conference on Object-Oriented Programming, Systems, Languages, and Applications, ACM Press, 2004, pp. 432–448.

[35] S. Person, M.B. Dwyer, S. Elbaum, C.S. Păsăreanu, Differential symbolic execution, in: Proceedings of the 16th ACM SIGSOFT International Symposium on Foundations of software engineering, SIGSOFT '08/FSE-16, ACM, New York, NY, USA, 2008, pp. 226–237.

[36] R. Santelices, M.J. Harrold, Probabilistic slicing for predictive impact analysis, Technical Report CERCS, GIT-CERCS-10-10, College of Computing, Georgia Institute of Technology, 2010.

[37] S. Horwitz, J. Prins, T. Reps, Integrating noninterfering versions of programs, ACM Trans. Program. Lang. Syst. 11 (3) (1989) 345–387.

[38] D.E. Perry, H.P. Siy, L.G. Votta, Parallel changes in large-scale software development: an observational case study, ACM Trans. Softw. Eng. Methodol. 10 (3) (2001) 308–337.

[39] S. Horwitz, Identifying the semantic and textual differences between two versions of a program, in: Proceedings of the ACM SIGPLAN 1990 Conference on Programming Language Design and Implementation, PLDI '90, ACM, New York, NY, USA, 1990, pp. 234–245.

[40] D. Jackson, D.A. Ladd, Semantic diff: a tool for summarizing the effects of modifications, in: Proceedings of the International Conference on Software Maintenance, ICSM '94, IEEE Computer Society, Washington, DC, USA, 1994, pp. 243–252.

[41] A. Loh, M. Kim, Lsdiff: a program differencing tool to identify systematic structural differences, in: Proceedings of the 32nd ACM/IEEE International Conference on Software Engineering, ICSE '10, vol. 2, ACM, New York, NY, USA, 2010, pp. 263–266.

[42] F.I. Vokolos, P.G. Frankl, Empirical evaluation of the textual differencing regression testing technique, in: Proceedings of the International Conference on Software Maintenance, ICSM '98, IEEE Computer Society, Washington, DC, USA, 1998, pp. 44.

[43] W. Miller, E.W. Myers, A file comparison program, Softw. Pract. Exper. 15 (11) (1985) 1025–1040.

[44] K. Zhang, D. Shasha, Simple fast algorithms for the editing distance between trees and related problems, SIAM J. Comput. 18 (6) (1989) 1245–1262.

[45] T. Apiwattanapong, A. Orso, M.J. Harrold, Jdiff: a differencing technique and tool for object-oriented programs, Autom. Softw. Eng. 14 (1) (2007) 3–36.

[46] D. Binkley, S. Horwitz, T. Reps, Program integration for languages with procedure calls, ACM Trans. Softw. Eng. Methodol. 4 (1995) 3–35.

[47] S. Horwitz, J. Prins, T. Reps, On the adequacy of program dependence graphs for representing programs, in: Proceedings of the 15th ACM SIGPLAN-SIGACT Symposium on Principles of Programming Languages, POPL '88, ACM, New York, NY, USA, 1988, pp. 146–157.

[48] J.M. Voas, Pie: a dynamic failure-based technique, IEEE Trans. Softw. Eng. 18 (1992) 717–727.

[49] S.K. Lahiri, C. Hawblitzel, M. Kawaguchi, H. Rebêlo, Symdiff: a language-agnostic semantic diff tool for imperative programs, in: Proceedings of the 24th International Conference on Computer Aided Verification, CAV'12, Springer-Verlag, Berlin, Heidelberg, 2012, pp. 712–717.

[50] R. Santelices, M.J Harrold, A. Orso, Precisely detecting runtime change interactions for evolving software, in: International Conference on Software Testing, Verification and Validation (ICST), IEEE, 2010.

[51] R. Santelices, P.K Chittimalli, T. Apiwattanapong, A. Orso, M.J. Harrold, Test-suite augmentation for evolving software, in: Proceedings of the 2008 23rd IEEE/ACM International Conference on Automated Software Engineering, ASE '08, IEEE Computer Society, Washington, DC, USA, 2008, pp. 218–227.

[52] A. Zeller, Yesterday, my program worked. today, it does not. why? in: Proceedings of the ESEC/FSE'99, 7th European Software Engineering Conference, Lecture Notes in Computer Science, vol. 1687, Springer, September 1999, pp. 253–267.

[53] W. Jin, A. Orso, T. Xie, Bert: a tool for behavioral regression testing, in: Proceedings of the 18th ACM SIGSOFT Symposium on the Foundations of Software Engineering (FSE 2010), Research Demonstration, November 2010, pp. 361–362.

[54] W. Jin, A. Orso, T. Xie, Automated behavioral regression testing, in: Proceedings of the 2010 Third International Conference on Software Testing, Verification and Validation, ICST '10, IEEE Computer Society, Washington, DC, USA, 2010, pp. 137–146.

[55] B. Korel, A.M. Al-Yami, Automated regression test generation, in: Proceedings of the 1998 ACM SIGSOFT International Symposium on Software Testing and Analysis, ISSTA '98, ACM, New York, NY, USA, 1998, pp. 143–152.

[56] E.J. Weyuker, On testing non-testable programs, Comput. J. 25 (4) (1982) 465–470.

[57] G. Fraser, N. Walkinshaw, Behaviourally adequate software testing, in: 2008 International Conference on Software Testing, Verification, and Validation, 2012, pp. 300–309.

[58] D. Qi, W. Sumner, F. Qin, M. Zheng, X. Zhang, A. Roychoudhury, Modeling software execution environment, in: 19th IEEE Working Conference on Reverse Engineering, WCRE'12, 2012.

[59] N. Rungta, E.G. Mercer, W. Visser, Efficient testing of concurrent programs with abstraction-guided symbolic execution, in: Proceedings of the 16th International SPIN Workshop on Model Checking Software, Springer-Verlag, Berlin, Heidelberg, 2009, pp. 174–191.

[60] C. Cadar, V. Ganesh, P.M. Pawlowski, D.L. Dill, D.R. Engler, Exe: automatically generating inputs of death, ACM Trans. Inf. Syst. Secur. 12 (2008) 10:1–10:38.

[61] M. Papadakis, N. Malevris, An empirical evaluation of the first and second order mutation testing strategies, in: Software Testing Verification and Validation Workshop, 2010, pp. 90–99.

[62] B. Korel, Automated software test data generation, IEEE Trans. Softw. Eng. 16 (8) (1990) 870–879.

[63] N.E. Beckman, A.V. Nori, S.K. Rajamani, R.J. Simmons, Proofs from tests, in: Proceedings of the 2008 International Symposium on Software Testing and Analysis, ISSTA '08, 2008, pp. 3–14.

[64] P. Godefroid, A.V. Nori, S.K. Rajamani, S.D. Tetali, Compositional may-must program analysis: unleashing the power of alternation, in: Proceedings of the 37th Annual ACM SIGPLAN-SIGACT Symposium on Principles of Programming Languages, POPL '10, 2010, pp. 43–56.

[65] A. Banerjee, A. Roychoudhury, J.A. Harlie, Z. Liang, Golden implementation driven software debugging, in: Proceedings of the 18th ACM SIGSOFT International Symposium on Foundations of Software Engineering, FSE '10, ACM, New York, NY, USA, 2010, pp. 177–186.

[66] M. Staats, M.W. Whalen, M.P. Heimdahl, Programs, tests, and oracles: the foundations of testing revisited, in: Proceedings of the 33rd International Conference on Software Engineering, ICSE '11, ACM, New York, NY, USA, 2011, pp. 391–400.

[67] H. Zhu, P.A.V. Hall, J.H.R. May, Software unit test coverage and adequacy, ACM Comput. Surv. 29 (1997) 366–427.

[68] W.E. Howden, Weak mutation testing and completeness of test sets, IEEE Trans. Softw. Eng. 8 (4) (982) 371–379.

[69] L. Morell, A theory of fault-based testing, IEEE Trans. Softw. Eng. 16 (1990) 844–857.

[70] M. Fisher II, J. Wloka, F. Tip, B.G. Ryder, A. Luchansky, An evaluation of change-based coverage criteria, in: Proceedings of the 10th ACM SIGPLAN-SIGSOFT Workshop on Program Analysis for Software Tools, PASTE '11, ACM, New York, NY, USA, 2011, pp. 21–28.

[71] P. Frankl, E. Weyuker, A formal analysis of the fault-detecting ability of testing methods, IEEE Trans. Softw. Eng. 19 (3) (1993) 202–213.

[72] S. Ntafos, A comparison of some structural testing strategies, IEEE Trans. Softw. Eng. 14 (6) (1988) 868–874.

[73] H. Zhu, A formal analysis of the subsume relation between software test adequacy criteria, IEEE Trans. Softw. Eng. 22 (4) (1996) 248–255.

[74] E.J. Weyuker, B. Jeng, Analyzing partition testing strategies, IEEE Trans. Softw. Eng. 17 (1991) 703–711.

[75] D. Hamlet, R. Taylor, Partition testing does not inspire confidence (program testing), IEEE Trans. Softw. Eng. 16 (1990) 1402–1411.

[76] J.W. Duran, S.C. Ntafos, An evaluation of random testing. IEEE Trans. Softw. Eng. 10 (4) (1984) 438–444.

[77] S. Ntafos, On random and partition testing, in: Proceedings of the 1998 ACM SIGSOFT International Symposium on Software Testing and Analysis, ISSTA '98, ACM, New York, NY, USA, 1998, pp. 42–48.

[78] M. Böhme, Software regression as change of input partitioning, in: Proceedings of the 2012 International Conference on Software Engineering, ICSE 2012, IEEE Press, Piscataway, NJ, USA, 2012, pp. 1523–1526.

[79] T.L. Graves, M.J. Harrold, J.M. Kim, A. Porter, G. Rothermel, An empirical study of regression test selection techniques, ACM Trans. Softw. Eng. Methodol. 10 (2001) 184–208.

[80] R. Gupta, M. Jean, M.J. Harrold, M.L. Soffa, An approach to regression testing using slicing, in: Proceedings of the Conference on Software Maintenance, IEEE Computer Society Press, 1992, pp. 299–308.

[81] Y.F. Chen, D.S. Rosenblum, K.P. Vo, Testtube: a system for selective regression testing, in: Proceedings of the 16th International Conference on Software Engineering, ICSE '94, IEEE Computer Society Press, Los Alamitos, CA, USA, 1994, pp. 211–220.

[82] M.J. Harrold, R. Gupta, M.L. Soffa, A methodology for controlling the size of a test suite, ACM Trans. Softw. Eng. Methodol. 2 (1993) 270–285.

[83] J. Jones, M. Harrold, Test-suite reduction and prioritization for modified condition/decision coverage, IEEE Trans. Softw. Eng. 29 (3) (2003) 195–209.

[84] G. Fraser, F. Wotawa, Redundancy based test-suite reduction, in: Proceedings of the 10th International Conference on Fundamental Approaches to Software Engineering, FASE'07, Springer-Verlag, Berlin, Heidelberg, 2007, pp. 291–305.

[85] G. Rothermel, M.J. Harrold, J. Ostrin, C. Hong, An empirical study of the effects of minimization on the fault detection capabilities of test suites, in: Proceedings of the International Conference on Software Maintenance, ICSM '98, IEEE Computer Society, Washington, DC, USA, 1998, 34–43.

[86] Y. Yu, J.A. Jones, M.J. Harrold, An empirical study of the effects of test-suite reduction on fault localization, in: Proceedings of the 30th International Conference on Software Engineering, ICSE '08, ACM, New York, NY, USA, 2008, pp. 201–210.

[87] S. McMaster, A.M. Memon, Fault detection probability analysis for coverage-based test suite reduction, in: ICSM, IEEE, 2007, pp. 335–344.

[88] D. Hao, L. Zhang, X. Wu, H. Mei, G. Rothermel, On-demand test suite reduction, in: Proceedings of the 2012 International Conference on Software Engineering, ICSE 2012, IEEE Press, Piscataway, NJ, USA, 2012, pp. 738–748.

[89] J.A. Jones, M.J. Harrold, Empirical evaluation of the tarantula automatic fault-localization technique, in: Proceedings of the 20th IEEE/ACM International Conference on Automated Software Engineering, ASE '05, ACM, New York, NY, USA, 2005, pp. 273–282.

[90] Z. Xu, Y. Kim, M. Kim, G. Rothermel, M.B. Cohen, Directed test suite augmentation: techniques and tradeoffs, in: Proceedings of the eighteenth ACM SIGSOFT International Symposium on Foundations of Software Engineering, FSE '10, ACM, New York, NY, USA, 2010, pp. 257–266.

[91] G. Fraser, A. Arcuri, Whole test suite generation, IEEE Trans. Softw. Eng. 99 (PrePrints) (2012).

[92] Z. Xu, G. Rothermel, Directed test suite augmentation, in: Proceedings of the 2009 16th Asia-Pacific Software Engineering Conference, APSEC '09, IEEE Computer Society, Washington, DC, USA, 2009, pp. 406–413.

[93] R.A. DeMillo, A.J. Offutt, Constraint-based automatic test data generation, IEEE Trans. Softw. Eng. 17 (9) (1991) 900–910.

[94] G. Fraser, A. Zeller, Mutation-driven generation of unit tests and oracles, IEEE Trans. Softw. Eng. 38 (2012) 278–292.

[95] D. Qi, A. Roychoudhury, Z. Liang, Test generation to expose changes in evolving programs, in: Proceedings of the IEEE/ACM International Conference on Automated Software Engineering, ASE '10, ACM, New York, NY, USA, 2010, pp. 397–406.

[96] K. Taneja, T. Xie, N. Tillmann, J. de Halleux, Express: guided path exploration for efficient regression test generation, in: ISSTA, ACM, 2011, pp. 1–11.

[97] G. Soares, R.Gheyi, T. Massoni, Automated behavioral testing of refactoring engines, IEEE Trans. Softw. Eng. 99 (PP) (2012) 19.

[98] P. McMinn, M. Harman, K. Lakhotia, Y. Hassoun, J. Wegener, Input domain reduction through irrelevant variable removal and its effect on local, global, and hybrid search-based structural test data generation, IEEE Trans. Softw. Eng. 38 (2012) 453–477.

[99] N .Tracey, J. Clark, K. Mander, J. McDermid, Automated test-data generation for exception conditions, Softw. Pract. Exper. 30 (1) (2000) 61–79.

[100] R. Ferguson, B. Korel, The chaining approach for software test data generation, ACM Trans. Softw. Eng. Methodol. 5 (1) (1996) 63–86.

[101] P. McMinn, M. Harman, D. Binkley, P. Tonella, The species per path approach to searchbased test data generation, in: Proceedings of the 2006 International Symposium on Software Testing and Analysis, ISSTA '06, ACM, New York, NY, USA, 2006, pp. 13–24.

[102] D. Kroening, A. Groce, E. Clarke, Counterexample guided abstraction refinement via program execution, in: Formal Methods and Software Engineering: 6th International Conference on Formal Engineering Methods, Springer, 2004, pp. 224–238.

[103] J. Strejček, M. Trtík, Abstracting path conditions, in: Proceedings of the 2012 International Symposium on Software Testing and Analysis, ISSTA 2012, ACM, New York, NY, USA, 2012, pp. 155–165.

[104] J.W. Duran, S.C. Ntafos, An evaluation of random testing, IEEE Trans. Softw. Eng. 10 (4) (1984) 438–444.

[105] C. Pacheco, M.D. Ernst, Randoop: feedback-directed random testing for java, in: Companion to the 22nd ACM SIGPLAN Conference on Object-Oriented Programming Systems and Applications Companion, OOPSLA '07, ACM, New York, NY, USA, 2007, pp. 815–816.

[106] A. Arcuri, M.Z.Z. Iqbal, L.C. Briand, Random testing: theoretical results and practical implications, IEEE Trans. Softw. Eng. 38 (2) (2012) 258–277.

[107] R. Santelices, M.J. Harrold, Applying aggressive propagation-based strategies for testing changes, in: Proceedings of the 2011 Fourth IEEE International Conference on Software Testing, Verification and Validation, ICST '11, IEEE Computer Society, Washington, DC, USA, 2011, pp. 11–20.

[108] D. Qi, A. Roychoudhury, Z. Liang, K. Vaswani, Darwin: an approach for debugging evolving programs, in: Proceedings of the the 7th Joint Meeting of the European Software Engineering Conference and the ACM SIGSOFT Symposium on The Foundations of Software Engineering, ESEC/FSE '09, ACM, New York, NY, USA, 2009, pp. 33–42.

[109] T. Apiwattanapong, R.A. Santelices, P.K. Chittimalli, A. Orso, M.J. Harrold, Matrix: maintenance-oriented testing requirement identifier and examiner, in: Proceedings of the Testing and Academic Industrial Conference Practice and Research Techniques (TAIC PART 2006), Windsor, UK, August 2006, pp. 137–146.

[110] Marcel Böhme, Bruno C.d.S. Oliveira, Abhik Roychoudhury, Partition-based regression verification, in: Proceedings of the 2013 International Conference on Software Engineering, ICSE 2013, San Francisco, CA, USA, 2013, pp. 1–10. <http://www.comp.nus.edu.sg/~mboehme/papers/ICSE13.pdf>.

ABOUT THE AUTHORS

Marcel Böhme received his Diplom (cf. M.Sc.) in Computer Science from Technische Universität Dresden, Germany. Advised by Dr. Roychoudhury since 2011, he is currently pursuing his doctoral studies in the area of Software Engineering at the National University of Singapore. Delving into the topic of Automated Test Suite Augmentation for Changed Behavior in Evolving Programs, his research interest includes automatic regression testing, fault and change correlation, impact analysis, semantic differencing, and semantic coverage criteria, amongst others. Generally, his research is driven towards establishing and extending the formal foundations of Software Testing and Debugging.

Abhik Roychoudhury is an Associate Professor of Computer Science at National University of Singapore, where he has been employed since 2001. Abhik received his Ph.D. in Computer Science from the State University of New York at Stony Brook in 2000. His research interests are in software testing and analysis with specific focus on software for real-time embedded systems. He has authored a book on "Embedded Systems and Software Validation" published by Elsevier (Morgan Kaufmann) Systems-on-Silicon series in 2009, which has been adopted for teaching at different universities. Abhik's research has led to scalable and usable analysis tools which enhance software quality as well as programmer productivity. Meaningful examples of such endeavor include the Chronos static analysis tool for ensuring time-predictable software execution, and the JSlice dynamic analysis tool for software debugging. Such tools have a substantial user-base spread across many different countries and have usage in teaching/development apart from research.

Dr. Bruno C.d.S. Oliveira received his M.Sc. in Computer Science from the Universidade do Minho (Portugal) in 2002, and his Ph.D. in Computer Science from the University of Oxford in 2007. In 2009 he joined Seoul National University as a Research Professor. He is currently a Senior Research Fellow in the National University of Singapore working in the area of Software Evolution and, in particular, on automated regression testing and program repair. He has previously conducted research on various aspects of Modularity in Functional Programming and Object-Oriented Programming languages, and has been a program committee member in several conferences and workshops in related areas.

CHAPTER THREE

Model Inference and Testing

Muhammad Naeem Irfan, Catherine Oriat, and Roland Groz

LIG, Computer Science Lab, Grenoble Institute of Technology, France

Contents

Abstract

For software systems, models can be learned from behavioral traces, available specifications, knowledge of experts, and other such sources. Software models help to steer testing and model checking of software systems. The model inference techniques extract structural and design information of a software system and present it as a formal model. This chapter briefly discusses the passive model inference and goes on to present the active model inference of software systems using the algorithm L^*. This algorithm switches between model inference and testing phases. In model inference phase it asks membership queries and records answers in a table to conjecture a model of a software system under inference. In testing phase it compares a conjectured model with the system under inference. If a test for a conjectured model fails, a counterexample is provided which helps to improve the conjectured model. Different counterexample processing methods are presented and analyzed to identify an efficient counterexample processing method. A counterexample processing method is said to be efficient if it helps to infer a model with fewer membership queries. An improved version of L^* is provided which avoids asking queries for some rows and columns of the table which helps to learn models with fewer queries.

1. INTRODUCTION

Prefabricated third-party components are used for rapid and cost effective development of software systems. Along performance, scalability, and functional requirements, software engineers also need to ensure other important properties like absence of deadlocks, chronological order of events, systems behave meaningfully in all provided circumstances. To avoid software failures or unexpected software behavior, system designers require checking all possible interactions between existing components and new system. They require software models or software specifications to understand possible behavior of components in new environment. Software development companies rarely provide complete software models due to different reasons that include copyright issues, models were not maintained, implementations evolved with the passage of time, etc.

A well-known solution to this problem is to infer a software model from source code [42]. But third-party components generally come as executable *black box* components, i.e., without source code. In the absence of source code the software models can be learned from interactions with software components. In literature we can find many *automata learning* algorithms [2, 37, 4, 23] that help to infer models of black box software components. The learning algorithms allow inferring finite state machine models by continuously interacting with black box components or from the *input/output* (*i/o*) traces obtained from interactions with components.

For inferring models of black box software systems we use the automata inference algorithm L^* proposed by Angluin [2] and some of its variants that focus on learning richer automata models with fewer tests. The algorithm L^* records the information in an observation table by interacting with a black box system and then conjectures a model from the observation table. The learning algorithm assumes that there is an oracle that knows the target model. The conjectured models are confirmed from the *oracle*. If a learned model is not correct the oracle replies with a *counterexample*. A counterexample is a string on the input set, which is accepted by the black box and refused by the conjecture or vice-versa. The learning algorithm learns a better model by taking this counterexample into account. The algorithm keeps on learning until the oracle cannot come up with a counterexample anymore. However, in the case of black box components such an oracle is not possible and this deficiency is alleviated by an approximate oracle [34, 24]. The counterexamples provided by approximate oracle are often very long, which results in increased number of queries. We have many counterexample processing methods [37, 31, 39, 24]. The chapter provides experiments that target on to identify an efficient counterexample processing method. The learning algorithms adds some unnecessary rows or columns to the observation table and the algorithm L_1 helps to avoid adding such rows and columns to the observation table which results in inferring models with reduced number of queries. Since for black box model inference searching and processing counterexamples is a complicated task, we have presented the Goodsplit algorithm [18] which learns the models of black box systems without requiring counterexamples.

Some basic definitions and notations commonly used throughout the chapter are provided in Section 2. The introduction to model inference and testing is provided in Section 3. The passive model learning techniques are briefly presented in Section 4. Section 5 provides the active learning using the model inference algorithm L^*. This section also presents: the variants of L^* algorithm that can learn richer automata models, different counterexample processing methods as they play a vital role for learning models with the algorithm L^*, the optimized learning algorithm L_1 which focuses on to learn by asking fewer queries, the learning algorithm GoodSplit which avoids searching and processing counterexamples, a concise introduction to learn the non–deterministic finite state automata models of black box components. Final section concludes the chapter.

2. DEFINITIONS AND NOTATIONS

This section provides the notations that we use to present definitions and algorithms formally. This section also provides the models that we use to model the behavior of software systems.

2.1 General Notations

Let Σ be a finite set or alphabet for a Deterministic Finite Automaton (DFA). The empty word is denoted by ϵ and the concatenation of two words s and e is expressed as $s \cdot e$ or se. The length of ω is the number of letters that a word ω contains and is denoted by $|\omega|$. If $\omega = u \cdot v$, then u and v are *prefix* and *suffix* of ω, respectively. Let Σ^+ be a sequence of letters constructed by taking one or more elements from Σ and Σ^* be a sequence of letters constructed by taking zero or more elements from Σ. Formally a word ω is a sequence of letters $a_1 a_2 \dots a_n \in \Sigma^*$ and the empty word is denoted by ϵ. The finite set of words over Σ of length exactly n is denoted as Σ^n, where n is a positive integer and $\Sigma^0 = \{\epsilon\}$. The finite set of words over Σ of length at most n and less than n are denoted as $\Sigma^{\leqslant n}$ and $\Sigma^{<n}$, respectively. Let $suffix^j(\omega)$ denote the *suffix* of a word ω of length j and $prefix^j(\omega)$ denote the *prefix* of the word ω of length j, where $j \in \{1, 2, \dots, |\omega|\}$. Let $prefixes(\omega)$ denote the set of all prefixes of ω and $suffixes(\omega)$ denotes the set of all suffixes of ω. Let $output^j(\omega)$ denote the output for jth input symbol in ω. For example, if we have an output 0010 for a word $aaba$ then $output^3(aaba) = 1$. A set is *prefix closed iff* all the prefixes of every element of the set are also elements of the set. A set is *suffix closed iff* all the suffixes of every element of the set are also elements of the set. The cardinality of a set D is denoted by $|D|$.

2.2 Finite State Machines

We use finite state machines (FSM) to design models of software systems. It is conceived as an abstract machine that can be in one of a finite number of states. A state is an instantaneous description of a system that captures the values of the variables at a particular instance of time. The state in which it is at any given time is called the current state. Typically a state is introduced when the system does not react the same way to the same trigger. It can change from one state to another when initiated by a triggering event or receiving an input, this is called a transition. A transition is a set of actions to be executed on receiving some input. A particular FSM is defined by a list of the possible transition states from each current state.

2.2.1 Deterministic Finite Automaton

The notion of DFA is formally defined as follows.

Definition 1. A DFA is a quintuple $(Q, \Sigma, \delta, F, q_0)$, where

- Q is the non–empty finite set of states,
- $q_0 \in Q$ is the initial state,
- Σ is the finite set of letters, i.e., the alphabet,
- $F \subseteq Q$ is the set of accepting/final states,
- $\delta : Q \times \Sigma \rightarrow Q$ is the transition function.

Initially a deterministic finite automaton is in the initial state q_0. From a current state the automaton uses the transition function δ to determine the next state. It reads a letter or a word of letters and using δ, it identifies a state in DFA, which can be accepting or non-accepting. The transition function for a word ω is extended as $\delta(q_0, \omega) = \delta(\ldots \delta(\delta(q_0, a_1), a_2) \ldots, a_n)$. A word ω is accepted by DFA iff $\delta(q_0, \omega) \in F$. We define an output function $\Lambda : Q \times \Sigma^* \rightarrow \{0, 1\}$, where $\Lambda(q_0, \omega) = 1$, if $\delta(q_0, \omega) \in F$, and $\Lambda(q_0, \omega) = 0$, otherwise.

2.2.2 Mealy Machine

The formal definition of a deterministic Mealy machine is given as follows.

Definition 2. A Mealy Machine is a sextuple $(Q, I, O, \delta, \lambda, q_0)$, where

- Q is the non–empty finite set of states,
- $q_0 \in Q$ is the initial state,
- I is the non–empty finite set of input symbols,
- O is the non–empty finite set of output symbols,
- $\delta : Q \times I \rightarrow Q$ is the transition function, which maps pairs of states and input symbols to the corresponding next states,
- $\lambda : Q \times I \rightarrow O$ is the output function, which maps pairs of states and input symbols to the corresponding output symbols.

Moreover, we assume $dom(\delta) = dom(\lambda) = Q \times I$, i.e., the Mealy machine is input enabled.

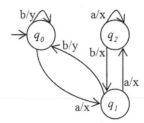

Fig. 1. Mealy machine.

For software systems all inputs may not be valid for every state. To make software systems input enabled, we introduce the output Ω for invalid inputs with transitions from the current state to itself. A Mealy machine on input set $I = \{a, b\}$ is presented in Fig. 1.

Initially the Mealy machine is in the initial state q_0. For a transition from a state $q \in Q$ an input $i \in I$ is read, output is produced by using the output function $\lambda(q, i)$ and the state reached is identified by using the transition function $\delta(q, i)$. When an input string ω composed of inputs $i_1 i_2 \ldots i_n \in I^+$ is provided to the Mealy machine, to calculate the state reached and output produced by the Mealy machine, both the transition and output functions are extended as follows. The transition function for ω is extended as $\delta(q_0, \omega) = \delta(\ldots \delta(\delta(q_0, i_1), i_2) \ldots, i_n)$. To extend output function for ω, i.e., $\lambda(q_0, \omega)$, the transition function is used to identify the state reached and the output function is used to calculate the output. The process is started by calculating $\lambda(q_0, i_1)$ and $\delta(q_0, i_1)$ from the initial state q_0, this takes to the next state $q_1 \in Q$. From q_1 output and next state are calculated by $\lambda(q_1, i_2)$ and $\delta(q_1, i_2)$, respectively. This process is continued until the state q_n is reached by $\lambda(q_{n-1}, i_n)$ and $\delta(q_{n-1}, i_n)$, and $i_1/o_1, i_2/o_2, \ldots i_n/o_n$ results in an output string $o_1 o_2 \ldots o_n$ of length n, where $n = |\omega|$. If the input string ω is provided to the Mealy machine and the state s_ω is reached then ω is the *access string* for s_ω. Let O^+ be a sequence of outputs constructed by taking one or more elements from O and O^* be a sequence of outputs constructed by taking zero or more elements from O.

3. TESTING BLACK BOX SYSTEMS

Testing is a process that manually or automatically verifies that the system conforms to the required specifications. It finds the differences between expected and system behavior. Generally, testing is done on a subset of all possible system behaviors, so it can find the presence of errors but not their absence. The field of combining automata learning and testing is accelerating with the passage of time. Automatically learning and verifying the software models has become a prominent approach for testing software systems. When software models are not available for black box software components, models can be learned from the interactions with components, available specifications, knowledge of the experts, and other such sources. The software models help to test and evaluate the test results. The models can be learned from traces of black box system behaviors. The passive learning techniques learn the models from a provided set of traces, whereas the active learning techniques iteratively interact with software systems to find out their behaviors. In the following sections, we discuss passive and active automata learning techniques in turn.

Testing is a process that manually or automatically verifies that a software system conforms to the required specifications. It finds the differences between a software system and its expected behavior. Generally, testing is done on a subset of all possible system behaviors, so it can find the presence of errors but not their absence. Automatically learning and verifying the software models has become a prominent approach for testing software systems. The software models help to steer the testing and evaluate the test results. When software models are not available for black box software components, models can be learned from the interactions with components, available specifications, knowledge of the experts, and other such sources. The software models can also be learned from traces of black box system behaviors.

Software model inference from implementations has become an important field in software testing [36, 32]. It has been used for the verification and testing of black box components [34, 38, 40]. It can be used to find the behavior of software systems where specifications are missing [33, 35]. Model inference has plenty of applications like program analysis [42], software testing [1], security testing [10, 21], dynamic testing [36], and integration testing [38]. Model inference approaches can mainly be divided into active and passive learning. Passive learning algorithms construct models from a fixed set of positive and negative examples, whereas active learning algorithms iteratively ask new queries during the model construction process. In the following sections, before discussing the active learning techniques in detail, we briefly present the passive learning techniques.

4. PASSIVE LEARNING

The passive learning algorithms learn the models from the provided set of positive traces, i.e., the set of traces which belong to the target language $L(DFA)$, and set of negative traces, i.e., the traces which do not belong to the target language or belong to $L(DFA)^C$. This technique binds to learn from the given data. These algorithms cannot perform tests by creating new tests and it is difficult to analyze how far one may be from the solution.

4.1 Inferring Models of Software Processes

Cook and Wolf [14] developed a technique termed as process discovery for the analysis of data describing the process events. They present three methods named as RNet, Ktail, and Markov, which range from purely algorithmic to purely statistical. They developed Markov method specifically for process discovery and adopted the other two methods from different domains. They automatically derive the formal model of a process from basic event data collected on the process [12]. Their approach views the process discovery as one of the grammatical inference problem. The data describing the behavior of the process are the sentences of some language and the grammar of that language is the formal model of the process. Cook et al. [13] implemented the proposed methods in a tool operating on process data sets. Although the proposed methods are automated, they still require the guidance from process engineers who are familiar with the considered process. The process engineers guide the methods for tuning the parameters built into them and for the selection and application of event data. The process models produced by the developed methods are initial models that can be refined by process engineers. The methods are developed for the model inference of software processes, however, they can be applied to other processes and behaviors. The methods were successful to discover the interprocess communication protocols at operating system level.

4.2 Generating Software Behavioral Models

One can learn the model for constraints on the data for the functional behavior as the boolean expressions [19] and the component interactions as the state machines [7, 14]. Both types of the models are important for testing and verifying different aspects of the software behavior. However, this type of models does not capture the interplay between data values and component interactions. To address this problem Lorenzoli et al. [29] presented the algorithm *GK-tail*, which automatically generates an extended finite state machines (*EFSMs*) from the software interaction strings. The algorithm *GK-tail* does not rely on the additional source of information like teachers and operates mainly as described in the following steps:

– merge the traces, which are input equivalent to create a unique trace annotated with multiple data values,

- generate the transition associated predicates with the help of *DAIKON* [19] from multiple data values,
- construct a tree-like structure *EFSM* from the interaction traces annotated with predicates,
- iteratively merge the states, which are equivalent to each other in *EFSM* to construct the final *EFSM*.

The input equivalent traces are different input traces, which invoke same sequence of methods in the target system and the equivalent transitions are the transitions, which have the same "tail." If the states cannot be distinguished by looking at the outgoing sequences, then they are said to have the same "tail." The method of merging the equivalent states is an extension to the algorithm *Ktail* by Biermann and Feldman [7], who consider a set of states to be equivalent, if all the member elements of the set of states are followed by the same paths. The sets of paths exiting a state can be infinite, thus comparing the states for all the possible exiting paths may be very expensive, so the length of the paths sequences to be compared is limited to k, exactly like Biermann et al. To evaluate *GK-tail*, Lorenzoli et al. implemented a prototype for Java programs based on, Aspectwerkz [3], theorem prover Simplify by Detlefs et al. [16], and *DAIKON*. To monitor systems and record the traces, this prototype is based on Aspectwerkz, to check for equivalence and implication between annotations on *Simplify* theorem prover and *DAIKON* for detecting the invariants to annotate the edges of *EFSM*. For the evaluation of *GK-tail* algorithm, *EFSM* models were generated for a set of open-source software applications, having different sizes and natures. The results by Lorenzoli et al. indicate that the size of the generated *EFSM* models does not depend on the size of the software implementations but on the size of interaction patterns within software components. They constructed *EFSMs* for a set of sample open-source applications of different size including Jabref [26] and jEdit [27], the calculated results show that the *EFSM* models were better and often more accurate than generating *FSMs* [7, 14] and learning the constraints [19] independently, especially for models of a non-trivial size.

5. ACTIVE LEARNING

For passive learning techniques, one is bound to learn a model from the provided set of traces. Such traces may not contain all the necessary information about the behavior of the system. It is not possible for the model inference techniques to learn a correct model from arbitrary set of traces. If the provided set of traces includes sufficient information about an implementation, i.e., what it can do and what it cannot, then inference techniques will be able to identify every state transition and distinguish all the non-equivalent states from each other. However, considering that the provided set of traces is unlikely to contain all the necessary information about target implementation, the learned model is a poor approximation of the real implementation. In effort to solve this problem a number of active learning techniques have been proposed. Active learning algorithms actively interact with the system to learn the models. Instead of relying on given traces, these algorithms can create and perform new tests on the target systems and find out how far they might be from the solution. The mathematical setting where the queries are asked to an oracle is called active learning. Active learning is used to construct the exact models of unknown black box systems in different communities. These techniques continuously interact with the target systems to produce the models. The learning algorithm L^* [2] is a well-known algorithm in this regard. The algorithm queries sequences of inputs to a black box system and organizes the replies in a table. On the basis of these replies a model is constructed, the conjectured model is iteratively improved with the help of a teacher (*oracle*). If the learned model is correct, the oracle replies "yes" or a *counterexample*, otherwise.

5.1 Angluin Learning Algorithm L^*

The learning algorithm L^* by Angluin [2] can learn the models of black box implementations. This algorithm assumes the formal model of black box implementations as unknown regular language, whose alphabet is known. The algorithm L^* infers the model of target black box system using the alphabet and inferred DFA is the exact behavior of black box. The algorithm queries sequences of inputs to the target black box and organizes the replies in a table. The rows and columns of the table are labeled with prefix closed and suffix closed sets of strings, respectively. This algorithm is based on two kinds of interactions with the black box: *membership* and *equivalence queries*. Making a membership query consists in giving to the black box a string that is constructed from the input set I and observing the answer, which can be recorded as 0 or "reject" and 1 or "accept." The membership queries are iteratively asked until conditions of compatibility and closure on the recorded answers are satisfied. The answers recorded into an observation table enable the algorithm to construct a conjecture, which is consistent with the answers. The algorithm then relies on the existence of a minimally adequate *teacher* or *oracle* that knows the unknown model. It asks an equivalence query to the oracle in order to determine whether the conjecture is equivalent to the black box or not. If the conjecture is not equivalent to the black box, the oracle replies with a counterexample or "yes," otherwise. By taking the counterexample into account, the algorithm iterates by asking new membership queries and constructing an improved conjecture, until we get an automaton that is equivalent to the black box.

5.1.1 Observation Table

The information collected by the algorithm as answers to the membership queries is organized in the observation table. Let $S \subseteq \Sigma^*$ be a *prefix closed* non-empty finite set, $E \subseteq \Sigma^*$ a *suffix closed* non-empty finite set, and T a finite function defined as $T : ((S \cup S \cdot \Sigma) \times E) \rightarrow \{0,1\}$. The observation table is a triple over the given alphabet Σ and is denoted as (S,E,T). The rows of the observation table are labeled with $S \cup S \cdot \Sigma$ and columns are labeled with E. For a row $s \in S \cup S \cdot \Sigma$ and column $e \in E$, the corresponding cell in the observation table is equal to $T(s,e)$. Now, $T(s,e)$ is "1," if $s \cdot e$ is accepted by the target model and "0," otherwise, i.e., $T(s,e) = \Lambda(q_0, s \cdot e)$. The observation table's rows S and columns E are non-empty and initially they contain ϵ, i.e., $S = E = \{\epsilon\}$. The algorithm runs by asking the membership and equivalence queries iteratively. Two rows $s_1, s_2 \in S \cup S \cdot \Sigma$ are said to be equivalent, *iff* $\forall e \in E, T(s_1, e) = T(s_2, e)$, and it is denoted as $s_1 \cong s_2$. For every row $s \in S \cup S \cdot \Sigma$, the equivalence class of row s is denoted by $[s]$. The observation table is finally used to construct a DFA conjecture. The rows labeled with strings from prefix closed set S are candidate states for DFA conjecture and columns labeled with strings from suffix closed set E are the sequences to distinguish these states. The elements $S \cdot \Sigma$ are used to build the transitions. An example of the observation table (S,E,T) for DFA learning is given in Table 1, where $\Sigma = \{a,b\}$.

To construct a DFA conjecture from an observation table, the table must satisfy two properties, closure and compatibility (in the original work the

Table 1 The observation table (S,E,T) for an automaton with alphabet $= \{a,b\}$.

		E
		ϵ
S	ϵ	
S · Σ	a	
	b	

compatibility concept is denoted as consistency). An observation table is closed if for each $s_1 \in S \cdot \Sigma$, there exists $s_2 \in S$ and $s_1 \cong s_2$. The observation table is compatible, if two rows $s_1, s_2 \in S$, and $s_1 \cong s_2$, then $s_1 \cdot a \cong s_2 \cdot a$, for $\forall a \in \Sigma$. To construct a conjecture that is consistent with the answers in (S, E, T), the table must be closed and compatible. If the observation table is not closed, then a possible state, which is present in the observation table may not appear in the conjecture. If the observation table is not compatible, then two states marked as equivalent in the observation table might be leading to two different states with same letter $a \in \Sigma$. In other words, if (S, E, T) is not compatible, then there exists $s_1, s_2 \in S$ and $s_1 \cong s_2$, and for some $a \in \Sigma, s_1 \cdot a \not\cong s_2 \cdot a$. When the observation table (S, E, T) satisfies the closure and compatibility properties, a DFA conjecture is build over the alphabet Σ as follows:

Definition 3.   Let the observation table (S, E, T) be closed and compatible, then DFA conjecture $Conj = (Q, \Sigma, \delta_D, F, q_0)$ is defined, where

- $Q = \{[s] | s \in S\}$,
- $q_0 = [\epsilon]$,
- $\delta([s], i) = [s \cdot i], \forall s \in S, i \in \Sigma$,
- $F = \{[s] | s \in S \wedge T(s, \epsilon) = 1\}$.

In order to verify that this conjecture is well-defined with respect to the observations recorded in the table (S, E, T), one can note that as S is a prefix closed non-empty set and it always contains ϵ, so q_0 is defined. Similarly as E is a non-empty suffix closed set, it also always contains ϵ. Thus, if $s_1, s_2 \in S$ and $[s_1] = [s_2]$ then $T(s_1) = T(s_1 \cdot \epsilon)$ and $T(s_2) = T(s_2 \cdot \epsilon)$ are defined and equal, which implies F is well-defined. To see that δ is well-defined, suppose two elements $s_1, s_2 \in S$ such that $[s_1] = [s_2]$. Since the observation table is compatible, $\forall a \in \Sigma, [s_1 \cdot a] = [s_2 \cdot a]$ and the observation table is also closed, so the rows $[s_1 \cdot a]$ and $[s_2 \cdot a]$ are equal to a common row $s \in S$. Hence, the conjecture is well-defined.

5.1.2 The Algorithm L*
The learning algorithm L^* organizes the gathered information into (S, E, T). It starts by initializing the rows S and columns E to $\{\epsilon\}$.

For all $s \in S$ and $e \in E$, the algorithm performs membership queries $s \cdot e$ and fills the cells of the observation table. Now, algorithm ensures that (S, E, T) is compatible and closed. If (S, E, T) is not compatible, incompatibility is resolved by finding two rows $s_1, s_2 \in S$, an alphabet letter $a \in \Sigma$ and a columns $e \in E$ such that $[s_1] = [s_2]$ but $T(s_1 \cdot a, e) \neq T(s_2 \cdot a, e)$, and adding the new suffix $a \cdot e$ to E. If (S, E, T) is not closed, the algorithm

---

Algorithm 1. The learning algorithm L^*

Input: Black box and the alphabet $\sum$
Output: DFA conjecture *Conj*
begin
    initialize the rows S and columns E of (S, E, T) to $\{\epsilon\}$ and
    $S \cdot \Sigma = \{\epsilon \cdot a\}, \forall a \in \Sigma$ complete (S, E, T) by asking membership queries $s \cdot e$
    such that $s \in (S \cup S \cdot \Sigma) \wedge e \in E$
    repeat
        while (S, E, T) is not closed or not compatible **do**
            if (S, E, T) *is not compatible* **then**
                find $s_1, s_2 \in S, a \in \Sigma$ and $e \in E$, such that
                $s_1 \cong s_2$, but $T(s_1 \cdot a, e) \neq T(s_2 \cdot a, e)$
                add $a \cdot e$ to E
                complete the table by asking membership queries for the column $a \cdot e$
            end
            if (S, E, T) *is not closed* **then**
                find $s_1 \in S \cdot \Sigma$ such that $s_1 \ncong s_2, \forall s_2 \in S$
                move the row s_1 to S
                add $s_1 \cdot a$ to $S \cdot \Sigma, \forall a \in \Sigma$
                complete the table by asking membership queries for new added rows
            end
        end
        construct the conjecture *Conj* from (S, E, T)
        ask the equivalence query to oracle for *Conj*
        if *oracle/teacher replies with a counterexample CE* **then**
            add *CE* and all the prefixes of *CE* to S
            complete the table by asking membership queries for new added rows
        end
    until *oracle replies "yes" to the conjecture Conj*;
    return the conjecture *Conj* from (S, E, T)
end

---

searches $s_1 \in S \cdot \Sigma$ such that $s_1 \ncong s_2$, for all $s_2 \in S$ and makes it closed by adding s_1 to S. This process is iterated until (S, E, T) is closed and compatible. The algorithm L^* eventually constructs the conjecture from (S, E, T) in accordance with Definition 3.

5.1.3 Complexity of L*

Angluin showed that the algorithm L^* can conjecture a minimum DFA in polynomial time on, the alphabet size $|\Sigma|$, the number of states in minimum conjecture n, and the length of the longest counterexample m. Initially S and E contain ϵ, each time (S, E, T) is discovered to be not compatible, one string is added to E. Since the observation table can be incompatible for at most $n - 1$ times, the total number of strings in E cannot exceed n. Each time (S, E, T) is discovered to be not closed one element is moved from $S \cdot \Sigma$ to S. This can happen for at most $n - 1$ times, and there can be at most $n - 1$ counterexamples. If the length of the longest counterexample CE provided by the oracle is m, for each counterexample at most m strings are added to S. Thus the total number of strings in S cannot exceed $n + m(n - 1)$. The worst case complexity of the algorithm in terms of membership queries is as follows:

$$(n + m(n - 1) + (n + m(n - 1))|\Sigma|)(n) = O(|\Sigma|mn^2).$$

5.2 DFA Inference of Mealy Models and Possible Optimizations

One bottleneck for the learning algorithm L^* is that it may require executing a lot of membership queries, although equivalence query is another tricky issue. Hungar et al. [23] proposed domain specific optimizations for reducing the number of membership queries for regular inference with the Angluin algorithm L^*. To reduce the number of membership queries, they introduced filters, which use properties like input determinism, prefix closure, independence, and symmetry of events. Most of the filters are linked to the specific structure of DFA, which enables the learning algorithm L^* to learn Mealy models. The reduction is done by having membership queries answered by the filters, in cases where the answer can be deduced from the membership queries of the so far accumulated knowledge of the system. They proposed the filters for testing reactive systems. They transform Mealy machines to DFA models with the help of Definition 4.

Definition 4. The transformation of a Mealy machine $\mathcal{M} = (Q_\mathcal{M}, I, O, \delta_\mathcal{M}, \lambda_\mathcal{M}, q_{0\mathcal{M}})$ to a DFA model $(Q, \Sigma, \delta, F, q_0)$ is defined as:

- $Q \supseteq Q_\mathcal{M} \cup \{q_{err}\}$, where q_{err} is a sink state (an artificial error state),
- $q_0 = q_{0\mathcal{M}}$,
- $\Sigma = I \cup O$,
- for each transition $\delta_\mathcal{M}(q, i) = q'$ where $i \in I, q, q' \in Q_\mathcal{M}$ and $\lambda_\mathcal{M}(q, i) = o$ where $o \in O$, the set Q contains the transient states $q_1 \cdot \ldots \cdot q_n$, with transitions;

  - $\delta(q, i) = q_1$ and $\delta(q_1, o) = q'$,

- the transitions $\delta(q, o) = q_{err}$ for all $q \in Q_\mathcal{M}$, and $o \in O$,
- the transitions $\delta(q_{err}, a) = q_{err}$ for all $a \in \Sigma$,
- $F = Q \backslash \{q_{err}\}$.

Mealy models are learned as DFA, which have only one sink state, i.e., all the rejecting strings lead to the sink state as described in Definition 4. Contrary to general languages, there is no switching from non-accepting to accepting states. For such languages, the optimization filters can be introduced that are described in the following. For qualitative evaluation of these filters, Hungar et al. [23] carried out experiments on four specific implementations of call center solutions. Each call center solution consists of telephone switch connected to a fixed number of telephones.

5.2.1 Prefix Closure

The filters suggest that once a string has been evaluated as accepting, then in the whole learning process, all the prefixes of the string will be evaluated to accepting without asking to the teacher. Similarly once a string has been evaluated as rejected, then all continuations of that string will be evaluated to rejected, i.e., all strings which have rejected string prefix to them will be evaluated to rejected without asking to the teacher. Thus the language to be learned is prefix closed. Formally these filters can be expressed as,

- for $\omega, \omega', \omega'' \in \Sigma^*$ and $\omega = \omega' \cdot \omega''$ and $T(\omega) = 1 \Rightarrow \Lambda(\omega') = $ *accept*,
- for $\omega, \omega' \in \Sigma^*$ and $\omega' \in prefix(\omega)$ and $T(\omega') = 0 \Rightarrow \Lambda(\omega) = $ *reject*.

For learning prefix closed languages of DFA models having single sink state, usage of such filters significantly reduces the number of membership queries. These two optimizations for a set of experiments conducted by Hungar et al. on scenarios for call center solution [23] gave a reduction of 72.22–77.85% membership queries.

5.2.2 Input Determinism

The second type of filters use the input determinism property, input determinism means that for an input sequence, we always have the same output or there is no non-determinism. These filters are proposed for the scenarios when learning reactive systems as DFA, and alphabet for learning such systems is union of input set I and output set O of the target system. The first filter of this type proposes that replacing just one output symbol in a word of an input deterministic language cannot be the word of the same language, for decomposition of ω, a word from the language $\mathcal{L}$, as $\omega = \omega' \cdot o \cdot \omega''$, where o belongs to O, then all the other words made by replacing o with other elements from alphabet $\Sigma \backslash \{o\}$ does not belong to the language $\mathcal{L}$ or in simple words all such words will be evaluated to rejected. The second filter of this type proposes that replacing one input symbol from

I with input symbol from O in a word of an input deterministic language cannot be the word of the same language. For decomposition of ω a word from the language $\mathcal{L}$ as $\omega = \omega' \cdot i \cdot \omega''$, where i belongs to I, then all the other words made by replacing i with elements from O do not belong to the language $\mathcal{L}$ or simply all such words will be rejected. Formally these filters can be expressed as,

- $\exists o \in O, a \in \Sigma \backslash \{o\}, \omega, \omega', \omega'' \in \Sigma^*$ such that $\omega = \omega' \cdot o \cdot \omega'' \wedge T(\omega' \cdot a \cdot \omega'') = 1 \wedge o \neq a \Rightarrow \Lambda(\omega) = reject,$
- $\exists i \in I, o \in O, \omega, \omega' \in \Sigma^*$ such that $\omega' \cdot o \in prefix(\omega) \wedge T(\omega' \cdot i) = 1 \Rightarrow \Lambda(\omega) = reject.$

The learning results for a set of experiments conducted on scenarios for call center solution [23] presented by Hungar et al. show that after introducing such filters the number of membership queries was reduced further from 50% up to 94.48%. This reduction is the additional reduction for the membership queries after the reduction by first type of filters.

5.2.3 Independence of Events

The intuition of independency in reactive systems can help to avoid asking unnecessary membership queries. Two devices in a system may require to perform actions that are independent of each other, i.e., the order of actions does not affect the results. If an input trace constructed from input traces executing two events independent of each other in two devices of the same system is accepted, then equivalence class of all traces to this trace is accepted. At first independent subparts of a trace with respect to the independence relation are identified, then shuffling these independent sub-traces in any order makes the equivalence class of a trace. The application of this filter along with first two types of filters further reduces the number of membership queries. The experiments on scenarios for call center solution [23] by Hungar et al. show that the reduction with such filters varies from 3.23% to 44.64%.

The experimental results [23] show that for the considered set of experiments on call center solutions, using all types of filters there was an overall reduction of 87.03–99.78%. For one of the examples initially 132, 340 membership queries were asked, whereas after the introduction of all the proposed filters it required only 289 membership queries, hence a reduction of 99.78%. Thus, modifying the learning algorithm by filtering out unnecessary queries enabled to perform quick and efficient learning. The approach of filters is quite flexible and is practical for the fast adaptations to different application domains.

5.3 Mealy Inference

The algorithm L^* can be adapted to learn reactive systems as Mealy machines [34, 28, 39] instead of DFA and this considerably improves the efficiency of the learning algorithm. DFA models require an intermediate state to model the i/o behavior of reactive systems whereas Mealy models do not. This is the reason why learning as DFA models require far more states to represent the same system as compared to Mealy models. DFA models lack the structure of i/o-based behavior; Mealy models are more succinct to represent reactive systems. Although, the algorithm L^* can learn Mealy models by model transformation techniques by taking the inputs I and outputs O of target unknown Mealy machine as alphabet for DFA as presented in Section 5.2. Alphabet set can be collection or product of the inputs and outputs, i.e., $\Sigma = I \cup O$ [23, 20] or $\Sigma = I \times O$ [30]. But this significantly increases size of alphabet and number of states in learned model, which increases the time complexity of the learning algorithm. The simpler way of handling this problem is to use the modified algorithm L^*, which learns Mealy machines directly.

Niese [34] proposed a Mealy adaptation of the algorithm L^* to learn reactive systems. He implemented the Mealy inference algorithm and conducted the experiments on four specific implementations of call center solutions. He observed that with the Mealy adaptation of the learning algorithm there was a noteworthy gain in terms of the membership queries and number of states. For instance, one of the examples for call center solutions discussed in Niese thesis and also presented in [23] required 132, 340 membership queries without any filter and with filters it required 289 membership queries, whereas Niese showed inferring the same example with Mealy inference technique required only 42 membership queries, which is an enormous reduction. For the number of states there was also a reduction of 60–90.12%.

The Mealy inference algorithm learns the target model by asking *output queries* [40]. For Mealy inference, the observation table is filled with output strings instead of *accept* or "1" and *reject* or "0." For the algorithm L^*, the columns of the observation table are initialized with ϵ, whereas the Mealy inference adaptation initializes the columns with the input set I. This change enables the algorithm to detect i/o to annotate the edges in the Mealy machines. The concept of closure and compatibility remains the same as for the algorithm L^*. However, to make the observation table closed and compatible, instead of comparing the boolean values, the outputs recorded in the table cells are compared. To process a counterexample CE, the prefixes of CE are added to the rows of observation table. To learn the models of black box software components, the Mealy inference algorithm can be used to infer the models with the following assumptions:

- input set I for the target machine is known,
- before each query the learner can always reset the target system to the initial state,
- the i/o interfaces of the machine are accessible, the interface from where an input can be sent is a input interface and the interface from where an output can be observed is an output interface.

5.3.1 The Mealy Inference Algorithm L_M^*

The Mealy machine inference algorithm L_M^* [34, 28, 39] learns the models as Mealy machines using the general settings of the algorithm L^*. The algorithm explores the target model by asking *output queries* [40] and organizes the outputs in the observation table. The output queries are iteratively asked until the observation table is closed and compatible. When the observation table is closed and compatible, a Mealy machine conjecture is constructed. The algorithm L_M^* then asks equivalence query to the oracle, if the conjectured model is not correct, the oracle replies with a counterexample. The algorithm processes the counterexample to improve the conjecture. If the oracle replies "yes," then the conjecture is correct and the algorithm terminates. We denote the Mealy machine to be learned by the algorithm L_M^* in Fig. 1 as $\mathcal{M} = (Q_\mathcal{M}, I, O, \delta_\mathcal{M}, \lambda_\mathcal{M}, q_{0\mathcal{M}})$. The observation table used by the algorithm L_M^* is described as follows.

Observation Table

The observation table contains the outputs from O^+ by the interactions with the target black box. The algorithm L_M^* sends the input strings from I^+ and records the outputs in the observation table. The observation table is defined as a triple (S_M, E_M, T_M), where $S_M \subseteq I^*$ is a prefix closed non-empty finite set, which labels the rows of the observation table, $E_M \subseteq I^+$ is a suffix closed non-empty finite set, which labels the columns of the observation table and T_M is a finite function, which maps $(S_M \cup S_M \cdot I) \times E_M$ to outputs O^+. The observation table rows S_M and columns E_M are non-empty and initially $S_M = \{\epsilon\}, S_M \cdot I = \{i \cdot \epsilon\}$ for all $i \in I$ and $E_M = I$. In the observation table, $\forall s \in S_M \cup S_M \cdot I, \forall e \in E, T_M(s, e) = suffix^{|e|}(\lambda_\mathcal{M}(q_{0\mathcal{M}}, s \cdot e))$. Since S_M is a prefix closed set, which includes ϵ and the columns of the observation table always contain the input set elements, this means that $\lambda_\mathcal{M}(q_{0\mathcal{M}}, s)$ can be calculated from the outputs already recorded in the observation table. Thus, recording the output suffix $suffix^{|e|}(\lambda_\mathcal{M}(q_{0\mathcal{M}}, s \cdot e))$ to the observation

Table 2 Initial observation table for the Mealy machine in Fig. 1.

		E_M	
		a	b
S_M	ϵ	x	y
$S_M \cdot I$	a	x	y
	b	x	y

table is sufficient. A word $\omega = s \cdot e$ is an input string or output query and on executing this output query the black box machine replies with $\lambda_{\mathcal{M}}(q_{0\mathcal{M}},\omega)$. However, in the observation table only $suffix^{|e|}(\lambda_{\mathcal{M}}(q_{0\mathcal{M}},\omega))$ is recorded. The initial observation table (S_M, E_M, T_M) for Mealy inference of the machine in Fig. 1 is presented in the Table 2, where the input set I has two elements $\{a,b\}$.

The equivalence of rows in the observation table is defined with the help of function T_M. Two rows $s_1, s_2 \in S_M \cup S_M \cdot I$ are said to be equivalent, iff $\forall e \in E_M, T_M(s_1,e) = T(s_2,e)$, and it is denoted as $s_1 \cong s_2$. For every row $s \in S_M \cup S_M \cdot I$, the equivalence class of row s is denoted by $[s]$. Like the algorithm L^*, to construct the conjecture, the algorithm L_M^* requires the observation table to satisfy the closure and compatibility properties. The observation table is closed, if $\forall s_1 \in S_M \cdot I$, there exists $s_2 \in S_M$ such that $s_1 \cong s_2$. The observation table is compatible whenever two rows $s_1 \cong s_2$ for $s_1, s_2 \in S_M$, then $s_1 \cdot i \cong s_2 \cdot i$ for $\forall i \in I$. On finding the observation table closed and compatible, the algorithm L_M^* eventually conjectures a Mealy machine. The rows labeled with strings from the prefix closed set S_M are the candidate states for the conjecture and the columns labeled with strings from suffix closed set E_M are the sequences to distinguish these states. The Mealy machine conjectured by the algorithm L_M^* is always a minimal machine.

Definition 5. Let (S_M, E_M, T_M) be a closed and compatible observation table, then the Mealy machine conjecture $M_M = (Q_M, I, O, \delta_M, \lambda_M, q_{0M})$ is defined, where

- $Q_M = \{[s] | s \in S_M\}$,
- $q_{0M} = [\epsilon]$,
- $\delta_M([s],i) = [s \cdot i], \forall s \in S_M, i \in I$,
- $\lambda_M([s],i) = T_M(s,i), \forall i \in I$.

We must show that M_M is a well-defined conjecture. Since S_M is non-empty prefix closed set and always contains ϵ, q_{0M} is well-defined. Suppose we have two elements $s_1, s_2 \in S_M$ such that $[s_1] = [s_2]$. Since the observation table is compatible, $\forall i \in I, [s_1 \cdot i] = [s_2 \cdot i]$ and since the observation table is also closed, the rows $[s_1 \cdot i]$ and $[s_2 \cdot i]$ are equal to a common row $s \in S_M$. Hence, δ_M is well-defined. Since E_M is non-empty and always contains inputs I, if there exists $s_1, s_2 \in S_M$ such that $s_1 \cong s_2$, then for all $i \in I$, we have $T_M(s_1, i) = T_M(s_2, i)$. Hence λ_M is also well-defined.

The Algorithm L_M^*

The learning algorithm L_M^* maintains the observation table (S_M, E_M, T_M).

Algorithm 2. The Algorithm $L_M{}^*$

Input: Black box and the input set I
Output: Mealy machine conjecture M_M
begin
    initialize the rows $S_M = \{\epsilon\}$, columns $E_M = I$ and $S_M \cdot I = \{\epsilon \cdot i\}, \forall i \in I$
    complete (S_M, E_M, T_M) by asking output queries $s \cdot e$ such that $s \in (S_M \cup S_M \cdot I)$
    $\wedge e \in E_M$
    repeat
        while (S_M, E_M, T_M) *is not closed or not compatible* **do**
            if (S_M, E_M, T_M) *is not compatible* **then**
                find $s_1, s_2 \in S_M, e \in E_M, i \in I$ such that $s_1 \cong s_2$, but
                $T_M(s_1 \cdot i, e) \neq T_M(s_2 \cdot i, e)$
                add $i \cdot e$ to E_M
                complete the table by asking output queries for the column $i \cdot e$
            end
            if (S_M, E_M, T_M) *is not closed* **then**
                find $s_1 \in S_M \cdot I$ such that $s_1 \not\cong s_2$ for all $s_2 \in S_M$
                move s_1 to S_M
                add $s_1 \cdot i$ to $S_M \cdot I$, for all $i \in I$
                complete the table by asking output queries for new added rows
            end
        end
        construct the conjecture M_M from (S_M, E_M, T_M)
        ask the equivalence query to oracle for M_M
        if *if oracle replies with a counterexample CE for M_M* **then**
            add all the prefixes of CE to S_M
            complete the table by asking output queries for new added rows
        end
    until *oracle replies "yes" to the conjecture M_M*;
    return the conjecture M_M from (S_M, E_M, T_M)
end

The set of rows S_M is initialized to $\{\epsilon\}$. The Mealy machine associates an output value with each transition edge, this output value is determined both by its current state and the input of the transition edge. The edges of

Mealy machines are annotated with i/o, where $i \in I$ and $o \in O$. Thus the set of columns E_M is initialized to I, it enables the algorithm to calculate the corresponding output label o for every $i \in I$ for all the transition edges in the conjectured model.

The output queries are constructed as $s \cdot e$, for all $s \in S_M \cup S_M \cdot I$ and $e \in E_M$, the observation table is completed by asking the output queries. The main loop of the algorithm L_M^* tests if (S_M, E_M, T_M) is closed and compatible. If (S_M, E_M, T_M) is not closed, then the algorithm L_M^* finds a row $s_1 \in S_M \cdot I$, such that $s_1 \not\cong s_2$, for all $s_2 \in S_M$. Then the algorithm L_M^* moves s_1 to S_M and completes the table. If (S_M, E_M, T_M) is not compatible, then the algorithm L_M^* finds $s_1, s_2 \in S_M, e \in E_M$, and $i \in I$ such that $s_1 \cong s_2$ but $T_M(s_1 \cdot i, e)$ is not equal to $T_M(s_2 \cdot i, e)$. Then the string $i \cdot e$ is added to E_M and T_M is extended to $(S_M \cup S_M \cdot I) \cdot (i \cdot e)$ by asking the output queries for missing elements. On finding the observation table (S_M, E_M, T_M) closed and compatible, the algorithm L_M^* builds the Mealy machine conjecture in accordance with Definition 5. The algorithm is explained with the help of the Mealy inference of the machine $\mathcal{M}$ in Fig. 1.

Example

The learning algorithm L_M^* begins by initializing (S_M, E_M, T_M). The rows S_M are initialized to $\{\epsilon\}$ and the columns E_M to $\{a, b\}$. The output queries are asked to complete the table. The initial observation table is shown in Table 2. This observation table is closed and compatible, thus the algorithm L_M^* conjectures the Mealy machine $M_{M'} = (Q_{M'}, I, O, \delta_{M'}, \lambda_{M'}, q_{0_M'})$ shown in Fig. 2.

Then, the algorithm L_M^* asks the equivalence query to the oracle. Since the conjectured model $M_{M'}$ is not correct, the oracle replies with a counterexample CE. There can be more than one counterexamples and oracle selects one from them. Let the counterexample CE selected by the oracle be $ababaab$, since

- $\lambda_{\mathcal{M}}(q_{0_\mathcal{M}}, ababaab) = xyxyxxx$, but
- $\lambda_{M'}(q_{0_M'}, ababaab) = xyxyxxy$.

a/x
b/y

$\rightarrow ([\epsilon])$

Fig. 2. The Mealy machine conjecture $M_{M'}$ from Table 2.

We have different methods to process the counterexamples, the processing of the counterexample CE with different methods is illustrated in Section 5.3.2.

Complexity of L_M^*

The learning algorithm L_M^* can conjecture a minimum Mealy machine in polynomial time on the input size $|I|$, the number of states in minimum conjecture n, and the length of the longest counterexample m. Initially E_M contains elements from input set I and its size is $|I|$, each time (S_M, E_M, T_M) is discovered to be not compatible, one string is added to E_M. Since the observation table can be incompatible for at most $n - 1$ times, the total number of strings in E_M cannot exceed $|I| + n - 1$. Initially S_M contains ϵ, i.e., one element. Each time the observation table (S_M, E_M, T_M) is discovered to be not closed, one element is moved from $S_M \cdot I$ to S_M. This can happen for at most $n - 1$ times, and there can be at most $n - 1$ counterexamples. If length of longest counterexample CE provided by the oracle is m, for each counterexample at most m strings are added to S. Thus the total number of strings in S cannot exceed $n + m(n - 1)$. The worst case complexity of the algorithm for number of output queries is described as follows:

$$(n + m(n - 1) + (n + m(n - 1))|I|)(|I| + n - 1) = O(|I|^2 mn + |I|mn^2).$$

5.3.2 Counterexamples

For black box model inference in general an equivalence oracle to reply about the correctness of learned models does not exist. A number of methods have been proposed to accommodate this deficiency; almost all of them involve a compromise on the precision. For inferring and testing black box software systems a common procedure is the random walk on inputs and providing the resulting input sequence in parallel to the conjectured model and black box to find the differences. Whenever a difference is found, the trace of all the inputs from the initial state to the last input resulting in output difference is considered as a counterexample. The counterexamples found by this method very often are non-optimal and there is a possibility that before reaching a state, where black box and conjectured model differ in behavior, many states are compared where the comparison was not required [24]. It is also not evident that one finds a counterexample at the very first attempt. Finding a counterexample may require a number of iterations. For certain implementations, it could be expensive and time consuming to calculate the outputs because of the internally involved delays.

Searching Counterexamples

In Angluin's black box learning framework the counterexamples help to iteratively refine the conjectured models. Angluin [2] proposed a random sampling oracle that selects a string x from input set I^+ according to some distribution and returns x with "yes" or "no," yes if x belongs to the language of the unknown model or no otherwise. All the calls to this oracle are independent of each other. This method may require a lot of strings x before finding a string, which can be taken as counterexample, i.e., a sequence that is accepted by the unknown model and rejected by the conjecture or vice-versa. Howar et al. [22] proposed the Evolving Hypothesis Blocking (*E.H. Blocking*) and Evolving Hypothesis Weighted (*E.H. Weighted*) algorithms, which steer the search of counterexamples. These algorithms generate the strings considering that every new string covers a different set of states in order to increase the probability of traversing states, which are not present in the conjectured model. For *E.H. Blocking* algorithm, the transitions are selected randomly from the $S \cdot I$ set, in a way that once the transition is used, it is excluded from the subsequent tests for the counterexample search. Continuing this process, when all the transitions of $S \cdot I$ set are disabled, then these are re-enabled. The algorithm *E.H. Weighted* requires annotating the conjecture transitions with a measure, which records a counter for the number of tests executed for a transition. This counter is reset whenever the transition changes. The algorithm uses these weights (counters) on all the transitions to select that transition for the counterexample search traversal. The probability of selecting a transition is inversely proportional to the increasing weight associated with the transition.

Balle [5] uses the algorithms $Balle_{L_1}$ and $Balle_{L_2}$ for learning unknown *DFA* models in the ZULU competition [11]. The difference between two algorithms lies in the method to search counterexamples. The algorithm $Balle_{L_1}$ uses the uniform distribution over alphabet for the counterexample search, whereas $Balle_{L_2}$ searches the counterexamples from *random walk* over the conjecture, with the probability of each transition depending on its destination's height. The search is based on the assumption that strings generated by traversing more transitions toward shorter leaves in the current conjecture are more likely to be counterexamples. In their implementation, they assign a weight to each transition using the expression,

$$ w\left(s,\sigma\right) = \left(\frac{1}{h_{\tau(s,\sigma)} - h_{\min} + 1} \right)^2, $$

where $h_{\tau(s,\sigma)}$ is the height of the leaf corresponding to the state $\tau(s,\sigma)$ and $h_{\min}$ is the height of the shortest leaf in the discrimination tree. Transition probabilities are obtained by normalizing these weights for each state: $p(s,\sigma) = w(s,\sigma)/W_s$ where $W_s = \Sigma_\sigma w(s,\sigma)$. They compute the transition probabilities for the hypothesis according to this rule. Even though they tried to make the algorithm $Balle_{L_2}$ efficient by introducing improvements described above and were expecting gain as compared to $Balle_{L_1}$ for learning the tasks offered by the ZULU challenge, however, they observed that both of the algorithms performed similarly and statically there was no significant difference.

Processing Counterexamples

The length of counterexamples is an important parameter to the complexity of L^* and its variant learning algorithms, which is evident from the section about the complexity of the algorithm L_M^*. The counterexample processing plays a vital role and directly affects the complexity. That is the reason why a significant number of different counterexample processing methods [37, 31, 39, 24] have been proposed.

The method to process counterexamples by Angluin [2] adds all prefixes of a counterexample to S_M, and two rows of S_M become equivalent only after processing the counterexamples. Rivest and Schapire [37] identified that incompatibilities in the observation table (S_M, E_M, T_M) can be avoided by keeping the rows S_M distinct. The compatibility condition requires that whenever two rows of S_M are equal, $s_1, s_2 \in S_M, s_1 \cong s_2$ then for $\forall i \in I$, $s_1 \cdot i \cong s_2 \cdot i$. But if the rows S_M are always distinct, that is for all $s_1, s_2 \in S_M$, always $s_1 \ncong s_2$, then the compatibility condition is trivially satisfied. The counterexample processing method by Rivest and Schapire adds only a single distinguishing string from a counterexample string CE to E_M. However, it may make up to $\log(m)$ output queries to find such a string, where $m = |CE|$. This method maintains the condition that all rows S_M of observation table are distinct, that is for all $s_1, s_2 \in S_M, s_1 \ncong s_2$.

The counterexample processing methods proposed by Maler and Pnueli [31], and Shahbaz and Groz [39] also add sequences only to columns E_M. Since for these methods S_M augments only when the observation table is not closed, this keeps the elements S_M distinct. Thus, always $|S_M| \leqslant n$, where n is the number of states in a conjectured model. We present the counterexample processing methods for the algorithm L_M^* in the following.

Counterexample Processing Algorithm by Angluin

The counterexample processing method by Angluin [2] requires to add all the prefixes of a counterexample to the rows of the observation table.

This method can be adapted for Mealy inference. If we have a counterexample CE, this method adds CE and all the prefixes of CE to S_M. Then the observation table is completed by extending T_M to $(S_M \cup S_M \cdot I) \cdot (E_M)$ by asking the output queries for new added rows. This method adapted for Mealy inference can be presented as Algorithm 3.

Algorithm 3. Counterexample processing by angluin for L_M^*

Input: Pre-refined observation table (S_M, E_M, T_M), CE
Output: Refined observation table (S_M, E_M, T_M)
begin
    for $j = 1$ to $|CE|$ **do**
        if $prefix^j (CE) \notin S_M$ **then**
            if $prefix^j(CE) \in S_M \cdot I$ **then**
                move the row $prefix^j(CE)$ to S_M
            end
        else
            add $prefix^j(CE)$ to S_M
        end
    end
    end
    construct the output queries for the new rows
    complete (S_M, E_M, T_M) by executing output queries
    make (S_M, E_M, T_M) closed and compatible
    return refined observation table (S_M, E_M, T_M)
end

After adding the prefixes of CE to S_M and completing the observation table, the table can be not closed or incompatible. The algorithm L_M^* makes the observation table closed and compatible, when both of these properties are satisfied, the algorithm conjectures the Mealy machine from the table.

Example
While learning the Mealy machine $\mathcal{M}$ in Fig. 1 with the algorithm L_M^*, for the initial conjecture $M_{M'}$ in Fig. 2, the oracle replies with a counterexample $CE = ababaab$. The counterexample processing method by Angluin requires to add CE and all of its prefixes to S_M that are not in S_M, i.e., $a, ab, aba, abab$, $ababa, ababaa, ababaab$ to the set S_M and the one letter extensions of these prefixes to $S_M \cdot I$ (that are not in $S_M \cup S_M \cdot I$), i.e., $aa, abb, abaa, ababb, ababab$, $ababaaa, ababaaba, ababaabb$ to the set $S_M \cdot I$. The function T_M is extended to $(S_M \cup S_M \cdot I) \cdot E_M$ by means of output queries for the missing entries. After

adding the prefixes of counterexample the observation table is presented in Table 3a.

The observation table in Table 3a is closed but not compatible. Since $\epsilon, ababa \in S_M, a \in I$ and $b \in E_M$ such that the underlined row labels $\epsilon \cong ababa$, but $T_M(\epsilon \cdot a, b)$ is not equal to $T_M(ababa \cdot a, b)$. Thus the algorithm L_M^* adds the string ab to E_M. The observation table is completed by executing the output queries for the newly added column ab. The observation table in Table 3b is closed and compatible, thus the algorithm L_M^* conjectures the Mealy machine shown in Fig. 3.

Table 3 The observation table in (a) after adding prefixes of *CE* is incompatible for rows with underlined labels. The observation table in (b) is made compatible by adding *ab* to E_M.

	a	b
<u>ε</u>	x	y
a	x	y
ab	x	y
aba	x	y
abab	x	y
<u>ababa</u>	x	y
ababaa	x	x
ababaab	x	y
b	x	y
aa	x	x
abb	x	y
abaa	x	x
ababb	x	y
ababab	x	y
ababaaa	x	x
ababaaba	x	x
ababaabb	x	y

(a) The observation table after adding prefixes of *CE*.

	a	b	ab
ε	x	y	xy
a	x	y	xx
ab	x	y	xy
aba	x	y	xx
abab	x	y	xy
ababa	x	y	xx
ababaa	x	x	xx
ababaab	x	y	xx
b	x	y	xy
aa	x	x	xx
abb	x	y	xy
abaa	x	x	xx
ababb	x	y	xy
ababab	x	y	xy
ababaaa	x	x	xx
ababaaba	x	x	xx
ababaabb	x	y	xy

(b) To make the observation table compatible *ab* is added to E_M.

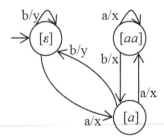

Fig. 3. Conjectured Mealy machine.

The conjectured model is correct, so the oracle replies "yes" and the algorithm L_M^* terminates. Thus a total of 51 output queries were asked to learn this example.

Counterexample Processing Algorithm by Rivest and Schapire

The counterexample processing method by Rivest and Schapire [37] adapted for L_M^* significantly improves the worst case number of output queries required to learn the Mealy machines. In the observation table (S_M, E_M, T_M), S_M is a prefix closed set representing the states of the conjecture. The counterexample processing method by Rivest and Schapire maintains the condition that all rows S_M of observation table are distinct, i.e., for all $s_1, s_2 \in S_M, s_1 \not\cong s_2$. Instead of adding prefixes of a counterexample CE to S_M, it adds only a single distinguishing string from CE to E_M. However, it may make up to $\log(m)$ output queries to find such a string, where m is the length of CE, i.e., $m = |CE|$. The counterexample processing method by Rivest and Schapire can be described as follows.

Let u_j be a sequence made up of first j actions in CE and v_j be a sequence made up of actions after first j actions, thus, $CE = u_j \cdot v_j$, where $0 \leqslant j \leqslant m$ such that $u_0 = v_m = \epsilon$ and $u_m = v_0 = CE$. Now, if we run u_j on a conjecture, the conjecture moves to some state q, where $q \in Q_M$. By construction this state q corresponds to a row $s \in S_M$. For the output query $s \cdot v_j$, let α_j be the output from a target Mealy machine $\mathcal{M}$ and β_j be the output from a conjecture. For a counterexample CE, we have $\alpha_0 \neq \beta_0$ and $\alpha_m = \beta_m$, and the point where $\alpha_j \neq \beta_j$ and $\alpha_{j+1} = \beta_{j+1}$ can be found using the binary search in $\log(m)$ output queries. Binary search can be performed by initializing j to $m/2$, if $\alpha_j \neq \beta_j$ then $j = 3(m/4)$ and $j = m/4$, otherwise. On finding such j, this method adds v_{j+1} to E_M. For $j < m$ and $i_j \in I$, the counterexample is $u_j \cdot i_j \cdot v_{j+1} = u_j \cdot v_j = CE$ and v_{j+1} distinguishes $s \cdot i_j$ from distinct rows of S_m, where $s \cdot i_j \in S_M \cdot I$. It is presented as Algorithm 4.

If L_M^* processes counterexamples with this method then size of S_M augments only when the observation table is not closed. Thus, we always have $|S_M| \leqslant n$, where n is the number of states in the conjecture. Since the compatibility condition requires that whenever two rows of S_M are equal, $s_1, s_2 \in S_M, s_1 \cong s_2$ then for $\forall i \in I$, we have $s_1 \cdot i \cong s_2 \cdot i$. To process counterexamples with Anguin's algorithm, L_M^* requires to add all prefixes of a counterexample to S_M and two rows of S_M become equivalent only after processing counterexamples. Since this counterexample processing method maintains the condition for all $s_1, s_2 \in S_M, s_1 \not\cong s_2$, compatibility condition is always trivially satisfied.

Algorithm 4. Counterexample Processing by Rivest and Schapire for $L_M{}^*$

Input: Pre-refined observation table (S_M, E_M, T_M), CE, Conjecture
Output: Refined observation table (S_M, E_M, T_M)
1: $CE = u_j \cdot v_j$, where u_j are first i actions and v_j are subsequent actions in CE
2: $q = \delta(q_{0M}, u_j)$ by running u_j on Conjecture, where $q \in Q_M$
3: Find the row $s \in S$, which corresponds to q
4: Calculate outputs α_j, β_j for output query $s \cdot v_j$ from target Mealy machine and learned Conjecture, respectively
5: By binary search find the point where $\alpha_j \neq \beta_j$ and $\alpha_{j+1} = \beta_{j+1}$
6: $u_j \cdot i_j \cdot v_{j+1} = u_j \cdot v_j = CE, j < m$
7: Add v_{j+1} to E_M
8: Construct the output queries for the new column
9: Complete (S_M, E_M, T_M) by executing output queries
10: Make (S_M, E_M, T_M) closed
11: **return** refined observation table (S_M, E_M, T_M)

Balcázar et al. pointed out that after processing a counterexample with this method the conjectured model may still classify the counterexample incorrectly, as another longer distinguish sequence from the same counterexample can improve the conjecture [4]. To address this, they propose to process distinguishing sequences from a counterexample until no further distinguishing sequence can be identified. This method adds only one suffix of a counterexample to E_M and requires a compromise on the suffix closure property for the observation table, consequently, the new conjecture may not be minimal and consistent with the observation table. The worst case complexity of L_M^* adapting this counterexample processing method in terms of output queries is $O(|I|^2 n + |I| n^2 + n \log(m))$.

Example

While learning the Mealy machine $\mathcal{M}$ in Fig. 1 with L_M^*, for the initial conjecture $M_M{}'$ in Fig. 2, the oracle replies with a counterexample $CE = ababaab$. The counterexample processing method by Rivest and Schapire searches the distinguishing sequence by means of binary search. The binary search divides $ababaab$ as $u_j \cdot v_j$, here $u_j = abab$ and $v_j = aab$. Now, by running $abab$ on the conjecture $M_M{}'$, the corresponding string from the observation table for the state reached is ϵ. The potential shorter counterexample sequence is $aab = \epsilon \cdot aab$. By running the output query aab on $\mathcal{M}$, we get

- $\lambda_{\mathcal{M}}(q_{0\mathcal{M}}, aab) = xxx$, but
- $\lambda_M{}'(q_{0M}{}', aab) = xxy$.

The search for shorter distinguishing sequence continues by selecting $u_j = ababaa$ and $v_j = b$, the string from the observation table for state reached is ϵ. New candidate sequence is $b = \epsilon \cdot b$, but

- $\lambda_{\mathcal{M}}(q_{0\mathcal{M}}, b) = x$, and
- $\lambda_{M'}(q_{0_M}', b) = x$.

The values for u_j and v_j are changed to $ababa$ and ab, respectively. New candidate sequence is $ab = \epsilon \cdot ab$, we get

- $\lambda_{\mathcal{M}}(q_{0\mathcal{M}}, ab) = xx$, and
- $\lambda_{M'}(q_{0_M}', ab) = xx$.

The binary search ends by finalizing $u_j = abab$ and $v_j = aab$. By running $abab$ on the conjecture $M_{M'}$, we get the access string ϵ. The row $\epsilon \cdot a$ along the column ab will make the table not closed. After adding the suffix $v_{j+1} = ab$, the observation table is presented in Table 4a.

From the observation table in Table 4a, it can be observed that the row $a \in S_M \cdot I$ is not equal to the only member ϵ of S_M, thus, the observation table is not closed. To make the table closed, the row a is moved to S_M and its one letter extensions are added to $S_M \cdot I$. Now, again the row aa of the observation table in Table 4b makes the table not closed. It is made closed by adding aa to S_M. The observation table in Table 4c is closed. Since with this counterexample processing method the observation table is always compatible, the Mealy machine shown in Fig. 3 is conjectured. The conjectured model is correct, so the oracle replies "yes" and the algorithm terminates. Thus, in total $24 = (21 + 3)$ output queries are asked.

Table 4 After adding the distinguishing suffix $v_{j+1} = ab$ to E_M, we get the observation table in (a). The underlined rows are the rows, which make the table not closed. The observation table in (c) is closed.

	a	b	ab
ϵ	x	y	xy
_a_	x	y	xx
b	x	y	xy

(a) Observation table after adding the distinguishing suffix $v_{j+1} = ab$.

	a	b	ab
ϵ	x	y	xy
a	x	y	xx
b	x	y	xy
_aa_	x	x	xx
ab	x	y	xy

(b) Make the observation table closed by moving the row a to S_M.

	a	b	ab
ϵ	x	y	xy
a	x	y	xx
aa	x	x	xx
b	x	y	xy
ab	x	y	xy
aaa	x	x	xx
aab	x	y	xx

(c) Make the observation table closed by moving aa to S_M.

Counterexample Processing Algorithm by Maler and Pnueli

The counterexample processing method by Maler and Pnueli [31] does not require any compromise on the suffix closure property of the observation table. If we have a counterexample CE, this method adds the counterexample CE and all the suffixes of CE to E_M that are not already a members of the set E_M. Since E_M always contains the set of inputs I, the algorithm adds only the suffixes of size $\geqslant 2$. Thus, the observation table is always suffix closed and the improved conjecture is always consistent to the observation table. Like Rivest and Schapire the size of distinct rows S_M augment only when a row from $S_M \cdot I$ is identified as distinct and is moved to S_M, thus, avoiding the incompatibilities trivially. The algorithm is presented as Algorithm 5.

Algorithm 5. Counterexample Processing by Maler and Pnueli for $L_M{}^*$

Input: Pre-refined observation table (S_M, E_M, T_M), CE
Output: Refined observation table (S_M, E_M, T_M)
begin
    for j = 2 to $|CE|$ **do**
        if *suffix^j*$(CE) \notin E_M$
            add *suffix^j*(CE) to E_M
        end
    end
    construct the output queries for the new columns
    complete (S_M, E_M, T_M) by executing output queries
    make (S_M, E_M, T_M) closed
    return refined observation table (S_M, E_M, T_M)
end

The observation table is completed for missing entries by executing the output queries. As S_M elements are always distinct, the observation table is compatible. If the observation table is not closed, it is made closed by finding $s_1 \in S_M \cdot I$ such that $s_1 \ncong s_2$ for all $s_2 \in S_M$ and then adding s_1 to S_M. On finding the table closed the Mealy machine is conjectured.

Example

While learning the Mealy machine $\mathcal{M}$ in Fig. 1a with L_M^*, for the initial conjecture $M_M{}'$ in Fig. 2, the oracle replies with a counterexample $CE = ababaab$. The counterexample processing method by Maler and Pnueli adds the counterexample $ababaab$ and elements of its prefixes set to the set E_M (the elements that are not already member of E_M), i.e., $ababaab$ and *babaab*, *abaab*, *baab*, *aab*, *ab* to columns of the observation table. The function T_M is extended to $(S_M \cup S_M \cdot I) \cdot E_M$ by means of output queries for the missing entries. After adding the suffixes of v, the observation table is presented in Table 5a.

From the observation table in Table 5a, it can be observed that the row $a \in S_M \cdot I$ is not equal to the only member ϵ of S_M, thus, the observation table is not closed. To make the table closed, the row a is moved to S_M and its one letter extensions are added to $S_M \cdot I$. Now, again the row aa of the observation table in Table 5b makes the table not closed. It is made closed by moving aa to S_M. The observation table in Table 5c is closed. Since with this counterexample processing method the observation table is always compatible, the Mealy machine shown in Fig. 3 is conjectured. The

Table 5 After adding *ababaab* and its suffixes *babaab*, *abaab*, *baab*, *aab*, *ab* to E_M, we get the observation table in (a). The underlined rows are the rows, which make the table not closed. The observation table in (b) is closed.

	a	b	ab	aab	baab	abaab	babaab	ababaab
ϵ	x	y	xy	xxx	yxxx	xyxxx	yxyxxx	xyxyxxx
$\underline{a}$	x	y	xx	xxx	yxxx	xxxxx	yxyxxx	xyxyxxx
b	x	y	xy	xxx	yxxx	xyxxx	yxyxxx	xyxyxxx

(a) Observation table after adding CE and its suffixes.

	a	b	ab	aab	baab	abaab	babaab	ababaab
ϵ	x	y	xy	xxx	yxxx	xyxxx	yxyxxx	xyxyxxx
a	x	y	xx	xxx	yxxx	xxxxx	yxyxxx	xyxyxxx
b	x	y	xy	xxx	yxxx	xyxxx	yxyxxx	xyxyxxx
$\underline{aa}$	x	x	xx	xxx	xxxxx	xxxxx	xxxxxx	xxxxxxx
ab	x	y	xy	xxx	yxxx	xyxxx	yxyxxx	xyxyxxx

(b) Make the observation table closed by moving the row a to S_M.

	a	b	ab	aab	baab	abaab	babaab	ababaab
ϵ	x	y	xy	xxx	yxxx	xyxxx	yxyxxx	xyxyxxx
a	x	y	xx	xxx	yxxx	xxxxx	yxyxxx	xyxyxxx
aa	x	x	xx	xxx	xxxx	xxxxx	xxxxxx	xxxxxxx
b	x	y	xy	xxx	yxxx	xyxxx	yxyxxx	xyxyxxx
ab	x	y	xy	xxx	yxxx	xyxxx	yxyxxx	xyxyxxx
aaa	x	x	xx	xxx	xxxx	xxxxx	xxxxxx	xxxxxxx
aab	x	y	xx	xxx	yxxx	xxxxx	yxyxxx	xyxyxxx

(c) Make the observation table closed by moving aa to S_M.

conjectured model is correct, so the oracle replies "yes" and the algorithm terminates. Thus, in total 56 output queries are asked.

Counterexample Processing Algorithm by Shahbaz and Groz

The counterexample processing method by Shahbaz and Groz [39] operates in the same manner as counterexample processing method by Maler and Pnueli [31]. The only difference is that this method drops the longest prefix of a counterexample CE that matches any element of $S_M \cup S_M \cdot I$, before adding the suffixes of the counterexample to E_M, whereas the method by Maler and Pnueli adds all the suffixes of counterexample to E_M. Thus for this method suffix closure property is trivially satisfied. As stated by Rivest and Schapire [37], incompatibilities can arise only when we have equivalent states in S_M, which can happen on adding counterexample prefixes to S_M. The counterexample processing algorithm by Shahbaz adds counterexample suffixes to E_M, and S_M augments only when the observation table is not closed, thus all the rows labeled by S_M elements are always distinct. After adding the suffixes to the observation table, it is required to make the table closed as it is always compatible. The algorithm is presented as Algorithm 6.

Algorithm 6. Counterexample Processing by Shahbaz

Input: Pre-refined observation table (S_M, E_M, T_M), CE
Output: Refined observation table (S_M, E_M, T_M)
begin
    divide CE as $u \cdot v$, where u is the longest prefix of CE such that $u \in (S_M \cup S_M \cdot I)$
    for $j = 2$ to $|v|$ **do**
        if $suffix^j(v) \notin E_M$ **then**
            add $suffix^j(v)$ to E_M
        end
    end
    construct the output queries for the new columns
    complete (S_M, E_M, T_M) by executing output queries
    make (S_M, E_M, T_M) closed
    return refined observation table (S_M, E_M, T_M)
end

The counterexample processing method by Shahbaz and Groz divides the CE as $u \cdot v$ where u is the longest prefix in $S_M \cup S_M \cdot I$. It adds all the suffixes of v to E_M that are not already member of the set E_M. Since E_M always contains the input set I, algorithm adds only the suffixes of size $\geqslant 2$.

Then the observation table is completed for missing entries. As S_M elements are always distinct the observation table is compatible. If the observation table is not closed, it is made closed by finding $s_1 \in S_M \cdot I$ such that $s_1 \not\cong s_2$ for all $s_2 \in S_M$ and then adding s_1 to S_M. On finding the table closed the Mealy machine is conjectured.

Example

While learning the Mealy machine $\mathcal{M}$ in Fig. 1 with the algorithm L_M^*, for the initial conjecture $M_M{}'$ in Fig. 2, the oracle replies with a counterexample $CE = ababaab$. The counterexample processing method by Shahbaz divides $ababaab$ as $u \cdot v$, where $u = a$ and $v = babaab$ ($a \in S_M \cup S_M \cdot I$ is the longest prefix of CE in the Table 2). Then it adds v and all the suffixes of v that are not in E_M, i.e., $babaab, abaab, baab, aab, ab$ to the set E_M. The function T_M is extended to $(S_M \cup S_M \cdot I) \cdot E_M$ by means of output queries for the missing entries. After adding the suffixes of v, the observation table is presented in Table 6a.

Table 6 After adding v and suffixes of v to E_M, we get the observation table in (a). The underlined rows are the rows, which make the table not closed. The observation table in (c) is closed.

	a	b	ab	aab	baab	abaab	babaab
ε	x	y	xy	xxx	yxxx	xyxxx	yxyxxx
_a_	x	y	xx	xxx	yxxx	xxxxx	yxyxxx
b	x	y	xy	xxx	yxxx	xyxxx	yxyxxx

(a) Observation table after adding suffixes of v.

	a	b	ab	aab	baab	abaab	babaab
ε	x	y	xy	xxx	yxxx	xyxxx	yxyxxx
a	x	y	xx	xxx	yxxx	xxxxx	yxyxxx
b	x	y	xy	xxx	yxxx	xyxxx	yxyxxx
_aa_	x	x	xx	xxx	xxxx	xxxxx	xxxxxx
ab	x	y	xy	xxx	yxxx	xyxxx	yxyxxx

(b) Make the observation table closed by moving the row a to S_M.

	a	b	ab	aab	baab	abaab	babaab
ε	x	y	xy	xxx	yxxx	xyxxx	yxyxxx
a	x	y	xx	xxx	yxxx	xxxxx	yxyxxx
aa	x	x	xx	xxx	xxxx	xxxxx	xxxxxx
b	x	y	xy	xxx	yxxx	xyxxx	yxyxxx
ab	x	y	xy	xxx	yxxx	xyxxx	yxyxxx
aaa	x	x	xx	xxx	xxxx	xxxxx	xxxxxx
aab	x	y	xx	xxx	yxxx	xxxxx	yxyxxx

(c) Make the observation table closed by moving aa to S_M.

From observation table in Table 6a, it can be observed that the row $a \in S_M \cdot I$ is not equal to the only member ϵ of S_M, thus the observation table is not closed. To make the table closed, the row a is moved to S_M and its one letter extensions are added to $S_M \cdot I$. Now, again the row aa of the observation table in Table 6b makes the table not closed. It is made closed by adding aa to S_M. The observation table in Table 6c is closed. Since with this counterexample processing method the observation table is always compatible, the Mealy machine shown in Fig. 3 is conjectured. The conjectured model is correct, the oracle replies "yes" and the algorithm terminates. Thus in total 49 output queries were asked.

Counterexample Processing Algorithm Suffix1by1

Like Rivest and Schapire [37], Maler and Pnueli [31], and Shahbaz and Groz [39], the Suffix1by1 counterexample processing method adds the suffixes of a counterexample to the columns E_M and the size of S_M augments only when the table is not closed. The Suffix1by1 method adheres to all the qualities of the counterexample processing methods presented previously. It adds the distinguishing sequences from a counterexample like Rivest and Schapire, it keeps the observation table suffix closed like Maler and Pnueli, and it removes the useless counterexample prefixes like Shahbaz and Groz.

If we have a counterexample CE, this method adds the suffixes one by increasing length from CE to E_M. Each time a suffix is added, the observation table is completed and the closure property is checked. The algorithm carries on this process until a suffix is found, which makes the table not closed and forces refinement. On finding such a suffix, this method stops adding the suffixes to E_M and conjectures the Mealy machine. The algorithm checks if CE is again a counterexample for the conjecture. If CE is again a counterexample, this means a longer suffix from the same counterexample CE can still improve the conjecture. This process is continued until CE can no longer help in improving the conjectured Mealy machine. This algorithm is described as Algorithm 7.

This method of processing the counterexamples also keeps the members S_M distinct, thus, while processing the counterexamples with this method, the observation table is always compatible. As the experiments show, in the case of non-optimal counterexamples, this method can have a dramatic impact. The rationale behind the proposed improvement in this method comes from the observation that random walks can cycle through states of the black box before reaching new states, which are not present in the learned model. Therefore, only the tail parts of such counterexamples actually correspond to

---

Algorithm 7. Counterexample Processing Suffix1by1

Input: Pre-refined observation table (S_M, E_M, T_M), CE
Output: Refined observation table (S_M, E_M, T_M)
begin
   while *CE is a counterexample* **do**
      for j = 2 to $|CE|$ **do**
         if *suffix^j(CE)* $\notin E_M$ **then**
            add *suffix^j(CE)* to E_M
            construct the output queries for the new columns
            complete (S_M, E_M, T_M) by executing output queries
            if (S_M, E_M, T_M) is not closed **then**
               break for loop
            end
         end
      end
      make (S_M, E_M, T_M) closed
      construct the conjecture M_M
   end
   return refined observation table (S_M, E_M, T_M)
end

---

discriminating sequences, thus considering suffixes of such counterexamples makes it possible to reduce the negative impact of unproductive cycles.

Example

While learning the Mealy machine $\mathcal{M}$ in Fig. 1 with the algorithm L_M^*, for the initial conjecture $M_M{'}$ in Fig. 2, the oracle replies with a counterexample $CE = ababaab$. The Suffix1by1 counterexample processing method adds the smallest suffix of CE to E_M that is not already a member of E_M. Since E_M is initialized with I, it starts by adding suffix of length 2, i.e., ab to the set E_M. The function T_M is extended to $(S_M \cup S_M \cdot I) \cdot E_M$ by means of output queries for the newly added column. After adding the suffix ab the observation table is presented in Table 7a.

From the observation table in Table 7a, it can be observed that the row $a \in S_M \cdot I$ is not equal to the only member ϵ of S_M, thus the observation table is not closed. To make the table closed, row a is moved to S_M and its one letter extensions are added to $S_M \cdot I$. Now, again the row aa of the observation table in Table 7b makes the table not closed. It is made closed by moving aa to S_M. The observation table in Table 7c is closed. This counterexample processing method always keeps the observation table compatible. So the

Table 7 We get the observation table in (a) after adding suffix ab to E_M. The underlined rows are the rows, which make the table not closed. The observation table in (c) is closed.

	a	b	ab
ε	x	y	xy
_a_	x	y	xx
b	x	y	xy

(a) The observation table after adding suffix ab.

	a	b	ab
ε	x	y	xy
a	x	y	xx
b	x	y	xy
_aa_	x	x	xx
ab	x	y	xy

(b) Make the observation table closed by moving the row a to S_M.

	a	b	ab
ε	x	y	xy
a	x	y	xx
aa	x	x	xx
b	x	y	xy
ab	x	y	xy
aaa	x	x	xx
aab	x	y	xx

(c) Make the table closed by moving the row aa to S_M.

Mealy machine shown in Fig. 3 is conjectured. The conjectured model is correct, so the oracle replies "yes" and algorithm terminates. Thus in total 21 output queries were asked.

Experiments

To analyze the insights of the algorithm L^*, Berg et al. [6] used randomly generated DFA examples. The random examples can be generated according to one's requirements ranging from simple to complex. We use randomly generated Mealy machines to analyze the algorithm L_M^* along the counterexample processing methods presented in this chapter. The random machine generator builds random Mealy machines with a provided number of inputs, outputs, and states. The FSM generation method is straightforward: we define Q the states of the machine and select an initial state from Q. Then, for each input $i \in I$, we select a random output from the given set of outputs $o \in O$ and select a random target state from states Q. The constructed Mealy machine serves as a black box to the learning algorithm. The generator simply takes as input: $|Q|, |I|, |O|$ and outputs a Mealy machine.

The oracle implementation compares a machine under inference and a learned conjecture to find the differences. The oracle implementation provides an input by selecting it randomly from the set of inputs, which is fed to a black box (machine under inference) and a conjecture. The output for each input is compared, as soon as a discrepancy is found, the sequence of inputs from beginning to the input which causes discrepancy is provided as a counterexample to the learning algorithm. However, it is possible that a search may fail to identify a discrepancy in reasonable time. For the experiments presented here, we set two bounds for counterexample search. We limit the maximum length for a counterexample sequence to five times the number of states in a machine under inference, if this limit is reached without finding a counterexample, then we reset the machine, reinitialize the search and start with a new sequence. The second bound is on the number of states in machine under inference, as soon as the number of states in a conjecture becomes equivalent to machine under inference, the learning process terminates. To increase our confidence for results calculated for a system under inference, the learning procedure is iterated for 30 times.

The set of random machines is generated with inputs $|I| = 2$, the outputs $|O| = 2$, and states $|Q| \in \{3, 4, \ldots, 40\}$. The algorithm L_M^* is executed to learn these machines repeatedly with the counterexample processing methods. The Fig. 4 presents the number of output queries asked by all of the methods. The vertical axis shows the number of output queries and horizontal axis shows the number of states. We can observe that by increasing the number of states, the gain with Rivest and Schapire method becomes more significant. For the smallest machine with State size $|Q| = 3$, the Rivest and Schapire, Suffix1by1, Angluin, Shahbaz and Groz, and Maler and Pnueli methods on the average requires $21, 21, 63, 42$, and 56 output queries, respectively. For the largest machine with $|Q| = 40$ requires $610, 775, 1287, 1215$, and 972 output queries, respectively. For the smallest machine all of the methods required only one counterexample, whereas for the largest machine the Rivest and Schapire, Suffix1by1, Angluin, Shahbaz and Groz, and Maler and

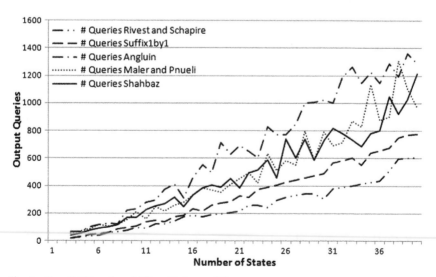

Fig. 4. $|I| = 2, |O| = 2$ and $|Q| \in \{3, 4, \ldots, 40\}$.

Pnueli methods requires $9, 5, 2.22, 2.75$, and 1.72 counterexamples respectively. The counterexample processing method by [37] performs better than the other counterexample processing methods. It adds only a distinguishing sequence from a counterexample CE and finds it in $\log{(m)}$ output queries. The compromise on suffix closure property by this method can result in a conjecture inconsistent to the observation table from which it is constructed. The Suffix1by1 is the second economic counterexample processing method and models conjectured with this method are always consistent with the observation table.

5.4 Improved Mealy Inference Algorithm

The Mealy inference algorithm L_1 [25] is an improved version of the learning algorithm L_M^*. The algorithm L_1 initially keeps the columns of the observation table empty with the intent to add only those elements from the set I, which are really required. But, Mealy adaptations [34, 28, 39] of the algorithm L^* initialize the columns with the input set I to calculate the annotations (labeling) for the transitions of conjectured models. To enable the algorithm L_1 to calculate the output labels for the transitions of the conjectured Mealy models, we record the output for the last input symbol of the access strings (input sequences) that label rows of the observation table. This output also helps to identify unnecessary rows of the observation table. The columns of the observation table are populated only to process the counterexamples, i.e., columns of the observation table contain only the distinguishing sequences, and sequences to keep the set of column labels suffix closed. To process the counterexample L_1 adds sequences from counterexample to columns of the observation table like Rivest and Schapire [37], Maler and Pnueli [31], and Shahbaz and Groz [39]. To process a counterexample, the algorithm L_1 adds the suffixes of the counterexample by increasing length to the columns of the observation table [24]. The observation table is populated with the outputs from O^+ that are calculated from the target system by sending input strings from I^+. The observation table used by the algorithm L_1 is described as follows.

5.4.1 Observation Table

The observation table is defined as a quadruple (S, E, L, T), where

- $S \subseteq I^*$ is a prefix closed non-empty finite set of access strings, which labels the rows of the observation table,
- $E \subseteq I^+$ is a suffix closed finite set, which labels the columns of the observation table,

Table 8 Initial table for mealy inference with L_1 of machine in Fig. 1.

		E
		Ø
S	ϵ	
$S \cdot I \backslash S$	a/x	
	b/y	

- for $S' = S \cup S \cdot I$, the finite function T maps $S' \times E$ to the set of non-empty output sequences $O^* \backslash \{\epsilon\}$ or O^+,
- and the finite function L maps $S' \backslash \{\epsilon\}$ to outputs O which are used to label the transitions.

The observation table rows S' are non-empty and initially, $S = \{\epsilon\}$ and $S \cdot I = I$. The output for the last input element for all the members of $S' \backslash \{\epsilon\}$ is recorded with the access strings S' by L. The columns E are initially $\emptyset$ and E augments only after processing the counterexamples. To process a counterexample, the suffixes of the counterexample are added to E by increasing length. The observation table is completed by extending T to $S' \cdot E$ by asking the output queries. The access strings are concatenated with the distinguishing sequences to construct the output queries as $s \cdot e$[1], for all $s \in S'$ and $e \in E$. In the observation table, $\forall s \in S', \forall e \in E, T(s,e) = suffix^{|e|}(\lambda(q_0, s \cdot e))$, and $L(s) = output^{|s|}(\lambda(q_0, s \cdot e))$. By means of function L, all the access strings $s \in S' \backslash \{\epsilon\}$ labeling the rows of the observation table contain the output o for the last input symbol of the access string s, and S' is prefix closed. This implies that $\lambda(q_0, s)$ can be calculated from the row labels. Thus, recording the suffix of the output query answer $suffix^{|e|}(\lambda(q_0, s \cdot e))$ to the cell labeled by row s and column e in the observation table is sufficient. Initial observation table (S, E, L, T) for Mealy inference of the machine in Fig. 1 is presented in Table 8, where the input set I is $\{a, b\}$.

The equivalence of rows in the observation table is defined with the help of function T. Two rows $s_1, s_2 \in S'$ are said to be equivalent, *iff* $\forall e \in E, T(s_1, e) = T(s_2, e)$, and it is denoted as $s_1 \cong s_2$. For every row $s \in S'$, the equivalence class of a row s is denoted by $[s]$. To construct the conjecture, the algorithm L_1 requires the observation table to satisfy the closure and

[1]The sequences $s \cdot e$ for $s \in S' \wedge e \in E$ and $s \cdot i$ for $s \in S' \wedge i \in I$ are constructed by considering the access strings only. The output for last input symbol of access strings recorded along rows $S' \backslash \{\epsilon\}$ of the observation table is used for mapping pair of a state and an input symbol to the corresponding output symbol (for annotating the transitions during conjecture construction) and identifying the valid access strings.

compatibility properties. The observation table is closed, if $\forall s_1 \in S \cdot I$, there exists $s_2 \in S$ such that $s_1 \cong s_2$. The observation table is compatible, if whenever two rows $s_1 \cong s_2$ for $s_1, s_2 \in S$, then $s_1 \cdot i \cong s_2 \cdot i$ for $\forall i \in I$. Since size of the rows S of observation table (S, E, L, T) increases only to make the table closed, the algorithm L_1 always maintains the condition that all rows S are distinct, i.e., for all $s_1, s_2 \in S, s_1 \not\cong s_2$. Thus, the compatibility condition is always trivially satisfied. On finding the observation table closed and compatible, the algorithm L_1 eventually conjectures a Mealy machine. The access strings S are the states for the conjecture and columns labeled with strings from suffix closed set E are the sequences that distinguish these states. The conjecture $Conj_1$ is defined as:

Definition 6. Let (S, E, L, T) be a closed and compatible observation table, then the Mealy machine conjecture $Conj_1 = (Q_C, I, O, \delta_C, \lambda_C, q_{0C})$ is defined, where

- $Q_C = \{[s] | s \in S\}$, (since $\forall s_1, s_2 \in S$, always $s_1 \not\cong s_2$ thus $|S| = |Q_C|$)
- $q_{0C} = [\epsilon], \epsilon \in S$ is the initial state of the conjecture
- $\delta_C([s], i) = [s \cdot i], \forall s \in S, i \in I$
- $\lambda_C([s], i) = L(s \cdot i), \forall s \in S, \forall i \in I \exists! s \cdot i \in S'$.

To verify that $Conj_1$ is a well-defined conjecture: since S is a non-empty prefix closed set and always contains ϵ, q_{0C} and Q_C are defined. For all $s \in S$ and $i \in I$, the string $s \cdot i$ is added to $S \cdot I$ exactly once. Thus, for every s and every i, there exists uniquely one $s \cdot i$ in S. Since the observation table is closed, the row $s \cdot i \cong s_1$ for some row $s_1 \in S$. Hence, δ_C is well-defined. Since every $s \cdot i$ is also associated with some output o by function L, λ_C is well-defined.

5.4.2 The Algorithm L_1

The learning algorithm L_1 maintains an observation table (S, E, L, T) to record the answers O^+ of the output queries I^+. The set of rows S is initialized to $\{\epsilon\}$ and the columns E at the beginning are $\emptyset$.

For the set of inputs I labeling rows $S \cdot I$, the outputs are calculated. All of the columns are empty, which implies that all the rows of the observation table are equivalent, i.e., $\forall s_1 \in S \cdot I$, there exists $s_2 \in S$, such that $s_1 \cong s_2$, thus, the observation table is closed. Since S has only one element ϵ, the compatibility condition is trivially satisfied. On finding the observation table closed and compatible, the algorithm L_1 conjectures a model (the initial conjecture is a single state "daisy" machine). For every access string $s \in S$ and for every input $i \in I$, there exists exactly one $s \cdot i \in S'$ and the output

Algorithm 8. The Algorithm L_1

Input: Black box and input set I
Output: Mealy Machine Conjecture
begin
   initialize the rows $S = \{\epsilon\}$, and columns $E = \emptyset$;
   execute output queries for $S \cdot I$ strings;
   since columns are empty, table is closed;
   construct a *conjecture* C;
   repeat
      search counterexamples;
      if *the oracle replies with a counterexample CE* **then**
         while *CE is a counterexample* **do**
            for $j = 1$ to $|CE|$ **do**
               if *suffix^j(CE)* $\notin E$ **then**
                  add *suffix^j(CE)* to E;
                  complete (S,E,T,L) by asking output queries $s \cdot e$ such that
                  $s \in S' \wedge e \in E$;
                  if (S,E,T,L) *is not closed* **then**
                     break for loop;
                  end
               end
            end
           while (S,E,L,T) *is not closed* **do**
               find $s_1 \in S \cdot I \backslash S$ such that $s_1 \not\cong s_2$, for all $s_2 \in S$;
               move s_1 to S;
               add $s_1 \cdot i$ to $S \cdot I$, for all $i \in I$;
               complete table by asking output queries for new added rows and
               find unnecessary rows;
            end
            construct a conjecture C;
         end
      end
   until *oracle replies yes to the conjecture C*;
end

for the last input i of the elements S' enables the algorithm L_1 to calculate the corresponding output label o for all the transitions of the conjectured model. Now the algorithm L_1 asks the equivalence query to the oracle. If the oracle replies with a counterexample CE, then L_1 adds one by one the suffixes by increasing length from CE to E. The observation table is completed by asking the output queries. After adding a suffix and completing the observation table, the closure property is checked. The algorithm continues adding suffixes until a suffix of CE is found, which makes the table not closed and forces refinement. On finding such a suffix, this method stops adding suffixes to E. Now the algorithm makes (S, E, L, T) closed by finding a row $s_1 \in S \cdot I$, such that $s_1 \not\cong s_2$, for all $s_2 \in S$ and moving s_1 to S. The one letter extensions of s_1 are added to $S \cdot I$. Since a row in the observation table corresponds to behavior of a state and Ω is output for a transition having same current and target state, on asking an output query for any of new added rows if the output found to be stored with access string of that row is Ω then it is marked as unnecessary row. On completing the table, again the closure property is checked. Since the observation table (S, E, L, T) is always compatible, on finding it closed, L_1 builds the Mealy machine conjecture in accordance with Definition 6.

Now the main while loop checks, if processed CE is again a counterexample for the learned conjecture. If "yes" (CE is still a counterexample), this means a longer suffix from the same counterexample CE can improve the conjecture. The algorithm L_1 relearns with CE and this process is continued until the counterexample CE can no longer help to improve the conjectured Mealy machine. After processing a counterexample and making the table closed, the algorithm again asks equivalence queries to the oracle. This process is continued until oracle is unable to find a counterexample and replies "yes" to the conjectured model. The algorithm is explained with the help of the Mealy inference of the machine $\mathcal{M}$ in Fig. 1.

5.4.3 Example

To illustrate the algorithm L_1, we learn the Mealy machine $\mathcal{M}$ given in Fig. 1. The algorithm L_1 begins by initializing (S, E, L, T), the rows S are initialized to $\{\epsilon\}$ and columns E to $\emptyset$. The output queries a and b are asked to find the outputs for input symbols $\{a, b\}$ labeling the rows $S \cdot I$. The initial observation table is shown in Table 8. This observation table has empty columns for all of the rows, thus, for every $s_2 \in S \cdot I$, there exists $s_1 \in S$ such that $s_2 \cong s_1$. Hence, the table is closed. Since for the algorithm L_1 the set of rows S always has distinct members, trivially the observation table is compatible.

On finding the observation table closed and compatible, the algorithm L_1 conjectures the model $Conj_1 = (Q_{C1}, I, O, \delta_{C1}, \lambda_{C1}, q_{0C1})$ which is similar to the conjecture shown in Fig. 2. To verify correctness of the one state machine conjecture, the algorithm L_1 asks an equivalence query to the oracle. The conjectured model $Conj_1$ is not correct, and there can be more than one counterexample and oracle replies with one from them.

- $\lambda(q_0, ababaab) = xyxyxxx$, but
- $\lambda_{C1}(q_{0C1}, ababaab) = xyxyxxy$.

Table 9 Model inference of Mealy machine in Fig. 3 with L_1.

	b
ϵ	y
a/x	y
b/y	y

(a) The observation table after adding the suffix b.

	b	ab
ϵ	y	xy
$\underline{a}/x$	y	xx
b/y	y	xy

(b) The observation table after adding the suffix ab.

	b	ab
ϵ	y	xy
a/x	y	xx
b/y	y	xy
$\underline{aa}/x$	x	xx
ab/y	y	xy

(c) Make the observation table closed by moving the row a to S.

	b	ab
ϵ	y	xy
a/x	y	xx
aa/x	x	xx
b/y	y	xy
ab/y	y	xy
aaa/x	x	xx
aab/x	y	xx

(d) Make the table closed by moving the row aa to S.

The oracle replies with the counterexample $ababaab/xyxyxxx$. The algorithm L_1 adds the smallest suffix b of the counterexample $ababaab$ to the columns E and completes the table, which remains closed as presented in Table 9a. Then suffix ab is added, which makes the observation table not closed as presented in Table 9b.

The row a as shown in Table 9b makes the observation table not closed. It is moved to S and its one letter extensions aa and ab are added to $S \cdot I$. To calculate the output for last input of access strings aa and ab, the algorithm does not require to execute the output queries separately. The output for the last input of access strings can be calculated from any of the output queries executed for that row. For instance if the algorithm asks the output query $ab \cdot ab$, the answer from the target machine is $xyxy$. The output for the last input string in ab is calculated as y and remaining xy is recorded in column ab of the observation table. After completing the table, it can be observed from Table 9c that the row aa makes the observation table not closed. Again the table is made closed by moving the row aa to S. The observation table in Table 9d is closed. Since the algorithm L_1 maintains the condition $\forall s_1, s_2 \in S$ always $s_1 \not\cong s_2$, the observation table is always compatible. Hence, L_1 conjectures the Mealy machine shown in Fig. 3.

The conjectured model is correct, the oracle replies "yes" and L_1 terminates. Initially two output queries were executed to calculate the outputs for $\{a,b\}$ labeling $S \cdot I$, thus, L_1 asks a total of 16 output queries to learn this example. Since all the inputs are valid for all of the states, none of the observation table row is marked as unnecessary row. Here, we have considered an example with a small set of inputs. If we learn a system with a large input set, the algorithm L_1 can restrict a bigger number of columns.

5.4.4 Complexity of L_1

The learning algorithm L_1 can conjecture a minimal Mealy machine in polynomial time on, the inputs size $|I|$, the number of states in the minimal conjecture n, and m the length of the longest distinguishing sequence added to the observation table (S, E, L, T) from a counterexample. Initially S contains ϵ, i.e., one element. Each time the observation table is discovered to be not closed, one element is moved from $S \cdot I \backslash S$ to S. This can happen for at most $n - 1$ times, hence, we always have $|S| \leqslant n$. The algorithm begins with columns $E = \emptyset$ and $|E| = 0$, each time to process a counterexample at most m suffixes can be added to E and at most there can be $n - 1$ counterexamples. Thus, the size of E cannot exceed $m(n - 1)$. Putting all this together the maximum cardinality of $S' \times E$ is $(n + n|I|) \times m(n - 1)$. The worst case complexity of L_1 in terms of output queries is $O(|I|mn^2)$. Since the algorithm avoids asking output queries for unnecessary rows (avoids asking output queries for observation table rows that have output Ω recorded with access strings), for every state the size of inputs $|I|$ is reduced to the size of the valid inputs.

5.5 Learning with Dictionaries

In the process of learning unknown models with the algorithm L^* or the algorithm L_M^*, we may very often repeatedly require the outputs for the output/membership queries that have already been asked or for the prefix of such queries. However, this new query might be for a different cell of the observation table. It consumes resources to execute the output queries in a black box and moreover it is useless to re-execute the query whose output is already known or can be inferred from the known results. This concept has been discussed for learning Mealy models as DFA in Section 5.2. The concept of dictionaries implementation is general and avoids asking the queries repeatedly. With the implementation of dictionaries, an output query will be executed in the black box only when its output cannot be inferred from the already executed queries. We can have certain variations for dictionary implementations, for instance the algorithm presented in Section 5.6.1 organizes accepted and rejected membership queries separately. At times the order of the output queries to be executed may also be important, executing longer queries before the smaller queries may help to answer smaller queries with the help of dictionaries.

5.6 Software Model Inference without Using Counterexamples

In literature, we can find a number of variants of the algorithm L^*, for all of these variants, the counterexamples play an important role [2, 37, 31, 38, 24]. For validating the conjectured model they require existence of an oracle. The existence of a precise and compact oracle for black box implementations is not possible. The absence of an oracle is alleviated by implementing an approximate oracle [34, 24, 25]. For black box testing and model inference, the techniques that do not require or require fewer counterexamples are preferred. In literature, we do not have many algorithms that have addressed this subject. The algorithm by Eisenstat et al. [18], named as GoodSplit, learns DFA models without using counterexamples. The GoodSplit algorithm alters structure of the observation table used by the algorithm L^* to record target model interactions. This learning algorithm adds all the possible distinguishing sequences to columns of the observation table taking into account the results by Trakhtenbrot et al. [41].

Domaratzki et al. [17] provided a lower bound on membership queries for identification of the distinct languages for DFA, they show that $(k - 1)$ $n \log_2(n) + O(n)$ bits are necessary and sufficient to specify a language accepted by automata with n states over a k symbols alphabet. Trakhtenbrot et al. indicate that a test set consisting of all strings of length at most about $\log_g \log_h(n)$ is sufficient to distinguish all distinct states in almost all complete finite state machines with n states, where g is the number of inputs and h is the number of outputs. For DFA, h is 2 as the result of the membership query can be *accept* or *reject*. For $d = \epsilon + \log_g \log_h(n)$, querying all suffixes of length $\leqslant d$ or all the suffixes in $I^{\leqslant d}$, for all the states and the states equivalent to these states entail about $k^{1+\epsilon} n \log_h(n)$ membership queries.

5.6.1 The Algorithm GoodSplit

The GoodSplit algorithm [18] also uses an observation table to record the outputs from target model interactions. This algorithm alters the structure of the observation table used for the algorithm L^*, it adds the distinguishing sequences to the columns of the observation table like [37, 31, 24, 25]. The models are learned as DFA accepting an unknown language over a given alphabet Σ. The outputs of the queries are recorded in the cells of the observation table in the form of *accept* or "1" and *reject* or "0," otherwise. This algorithm does not fill the observation table completely; some cells contain either *accept* or *reject*, depending on the output of the corresponding query, while other cells do not contain anything at all, until queries for those cells are not calculated. The rows of the observation table are states and the columns are the identifying sequences. The set of columns is denoted by $\Sigma^{\leqslant l}$, where l is the length of the longest suffix added to the observation table. The set of distinct rows is denoted by D and it is initialized with ϵ, where ϵ is the empty string. The set of all distinct and equivalent to distinct rows is denoted by P, i.e., $P = D \cup D \cdot \Sigma$. Two rows $r_1, r_2 \in P$ are said to be consistent *iff* $\forall e \in \Sigma^{\leqslant l}$, if both $r_1 \cdot e$ and $r_2 \cdot e$ are non-empty, then they must be equal, and it is denoted by $r_1 \cong_c r_2$. The inconsistency of rows r_1 and r_2 is denoted as $r_1 \ncong_c r_2$. For any row r the set D_r denotes the rows from D consistent with r. A row is identified as distinct, if it is not consistent with any of the distinct rows, in other words, two distinct rows are always inconsistent. For all $r \in P, [r]$ denotes the equivalence class of row r, the equivalence class of rows includes all the rows that are consistent with r. Initially the observation table has one row and one column both consisting of the empty string ϵ. The output for the only cell of the observation table is calculated and recorded. The set of executed queries is cached and consulted before asking new queries and is maintained as $A = A_0 \cup A_1$, where A_0 are the queries answered "0" and A_1 are the queries answered "1." The algorithm was proposed for the ZULU competition [11], where a limited number of queries were allowed. This algorithm uses the query limit as the termination criteria. The algorithm operates in the following steps,

1. For the transitions of the states, $\forall r \in D$ and $a \in \Sigma$, if $r \cdot a \notin P$, then add $r \cdot a$ to P and complete the table by asking membership queries.
2. If $\exists r_1 \in (P - D)$ and $\forall r_2 \in D, r_1 \ncong_c r_2$, then move r_1 to D. Repeat this process until P does not change anymore.
3. $\forall r \in (P - D)$, as long as $|D_r| > 1$, the algorithm selects greedily a suffix $e \in \Sigma^{\leqslant l}$ and queries $r \cdot e$. The greedy choice is made as: For $b \in \{0,1\}$ let $v_b(r,e) = |\{r' \in D_r : r' \cdot e \in A_b\}|$ that is the number of $r' \in D_r$ such that $r' \cdot t$ has been queried and answered b. Then $t \in \Sigma^{\leqslant l}$ is chosen to maximize $v(r,e) = min\{v_0(r,e), v_1(r,e)\}$.
4. The algorithm uses the following heuristic to decide whether or not to increment the current suffix length l. If more than 90% of table cells are filled, then l is incremented by 1, i.e., $(r,e) \in (P - D) \times \Sigma^{\leqslant l}$ and $r \cdot e \in A$ is greater than 90% then increment l.

5. For $\lceil |D|/2 \rceil$ random choices $(r,e) \in (P-D) \times \Sigma^{\leqslant l}$ such that output for $r \cdot e$ is not known, query both $r \cdot e$ and $r' \cdot e$ such that $r' \in D$ is consistent with r. Return to step 1.

After reaching the query limit, the algorithm executes step 1 and step 2, as if at this stage there are distinct states in $(P-D)$, they can be moved to D. The conjecture is constructed from the observation table as follows:

Definition 7. GoodSplit algorithm constructs a DFA model from the observation table and conjecture $Conj = (Q, \Sigma, \delta, F, q_0)$ is defined as,

- $Q = \{[r] | r \in D\}$
- $q_0 = [\epsilon]$
- $F = \{[r] | r \in D \wedge r \in A_1\}, \forall r \in D,$
  if $r \notin A$, then randomly select a label for r from Σ
- $\delta([r], a) = [r \cdot a], \forall r \in D, a \in \Sigma,$
  if more than one $r' \in D$ are consistent with $r \cdot a$, then one is selected randomly.

The strings in D are the distinct states for the inferred model and the initial state is ϵ. If the corresponding string of the state in the observation table belongs to A_1, then it is accepting state, and if it belongs to A_0, then it is a rejecting state. If the string does not belong to A, or the string has not been queried, then it is selected from a uniform distribution over *accept* or *reject*. For all $r \in D$, the transition function $\delta(r, a)$ maps the transition for state r with alphabet a to $r \cdot a$, if r is in D. Otherwise, $r' \in D$ is selected such that $r' \cong r \cdot a$, if $r \cdot a$ is consistent with more than one r', then from such r', one distinct state is selected randomly.

5.7 Learning the NFA Models

In many testing and verification applications the learned non–deterministic finite state automata (NFA) models are exponentially smaller than the corresponding DFA models. So there is a significant gain, when we use learning algorithms that learn the models as succinct NFA instead of DFA. But the issue is that class of NFA lacks the significant properties that most of the learning algorithms available in the literature use. For a given regular language there is no minimally unique NFA, so it is not clear which of the automaton is required to be learned. Denis et al. [15] introduced a subclass of NFA, the residual finite state automata (RFSA), this class shares the important properties with DFA class. For example there is unique minimal canonical RFSA for all the regular languages accepting them which is more succinct than the corresponding DFA. Bollig et al. [8, 9] proposed an active learning algorithm NL^* which learns the RFSA models. This algorithm alters the Angluin algorithm L^* to learn the RFSA models using the membership and equivalence queries. The algorithm NL^* requires $O(n^2)$ equivalence queries and mn^3 membership queries where n is the minimal number states in the target model and m is the length of the longest counterexample provided by the oracle. They introduced the concepts of *Composed and Prime Rows* to redefine closure and compatibility as *RFSA-Closure* and *RFSA-Compatibility*, respectively. They implemented the algorithm and used it to learn the regular languages described by the regular expressions. For considered set of experiments in most of the cases they observed that they required very few membership and equivalence queries as compared to L^*, and the RFSA learned by NL^* were much smaller than the corresponding DFA learned by L^*. Yokomori [43] has also shown that a specific class of NFA, called polynomially deterministic, is learnable in polynomial time using membership and equivalence queries.

6. CONCLUSION

This chapter presents the model inference and testing techniques for black box components. It provides details on how model inference can be used for testing. The algorithm L^* is presented formally along complexity discussion on number of membership queries. The algorithm L^* can be used to learn Mealy models as DFA, and DFA inference of Mealy models can be optimized with the help of filters. However, the direct Mealy inference algorithm L_M^* requires less number of queries than optimized DFA inference of Mealy models. The processing of counterexamples plays a vital role for model inference, thus efficient counterexamples processing method adapted for the learning algorithms significantly reduces the required number of output queries. We conducted experiments which show that the Suffix1by1 counterexample processing method is better than the others.

The algorithm L_M^* initializes the observation table columns with the input set I. The algorithm L_1 does not initialize the columns of the observation table with I and adds only those elements of I to the columns which are distinguishing sequences or suffixes of distinguishing sequences. The algorithm L_1 also identifies the unnecessary rows of the observation table and avoids asking output queries for them. Keeping fewer columns for the observation table and avoiding asking output queries for unnecessary rows results in asking fewer output queries. Since to learn black box models searching and processing counterexamples can be very expensive, the learning algorithm

GoodSplit provides a choice to learn black box models without using counterexamples and by taking into account some bounds on the length of distinguishing sequences. For non-deterministic models there is a gain when we use learning algorithms which learn the models as succinct NFA instead of DFA.

REFERENCES

[1] Fides Aarts, Bengt Jonsson, Johan Uijen, Generating models of infinite-state communication protocols using regular inference with abstraction, in: ICTSS, 2010, pp. 188–204.

[2] Dana Angluin, Learning regular sets from queries and counterexamples, Inf. Comput. 75 (2) (1987) 87–106.

[3] Aspectwerkz, 2007. <http://aspectwerkz.codehaus.org/>.

[4] José L. Balcázar, Josep Díaz, Ricard Gavaldá, Algorithms for learning finite automata from queries: a unified view, in: Advances in Algorithms, Languages, and Complexity, 1997, pp. 53–72.

[5] Borja Balle, Implementing kearns-vazirani algorithm for learning DFA only with membership queries, in: ZULU Workshop Organised During ICGI, 2010.

[6] Therese Berg, Bengt Jonsson, Martin Leucker, Mayank Saksena, Insights to angluin's learning, Electr. Notes Theor. Comput. Sci. 118 (2005) 3–18.

[7] A.W. Biermann, J.A. Feldman, On the synthesis of finite-state machines from samples of their behaviour, IEEE Trans. Comput. 21 (1972) 591–597.

[8] Benedikt Bollig, Peter Habermehl, Carsten Kern, Martin Leucker, Angluin-style learning of NFA, Research Report LSV-08-28, Laboratoire Spécification et Vérification, ENS Cachan, France, October 2008, p. 30.

[9] Benedikt Bollig, Peter Habermehl, Carsten Kern, Martin Leucker, Angluin-style learning of NFA, in: IJCAI, 2009, pp. 1004–1009.

[10] Chia Yuan Cho, Domagoj Babic, Eui Chul Richard Shin, Dawn Song, Inference and analysis of formal models of botnet command and control protocols, in: ACM Conference on Computer and Communications Security, 2010, pp. 426–439.

[11] David Combe, Myrtille Ponge, Colin de la Higuera, Jean-Christophe Janodet, Zulu: an interactive learning competition, in: ZULU Workshop Organised during ICGI, 2010.

[12] Jonathan E. Cook, Alexander L. Wolf, Automating process discovery through event-data analysis, in: ICSE, 1995, pp. 73–82.

[13] Jonathan E. Cook, Artur Klauser, Alexander L. Wolf, Benjamin G. Zorn, Semi-automatic, self-adaptive control of garbage collection rates in object databases, in: SIGMOD Conference, 1996, pp. 377–388.

[14] Jonathan E. Cook, Alexander L. Wolf, Discovering models of software processes from event-based data, ACM Trans. Softw. Eng. Methodol. 7 (3) (1998) 215–249.

[15] François Denis, Aurélien Lemay, Alain Terlutte, Residual finite state automata, Fundam. Inform. 51 (4) (2002) 339–368.

[16] David Detlefs, Greg Nelson, James B. Saxe, Simplify: a theorem prover for program checking, J. ACM 52(3) (2005) 365–473.

[17] Michael Domaratzki, Derek Kisman, Jeffrey Shallit, On the number of distinct languages accepted by finite automata with n states, J. Autom. Lang. Combin. 7 (4) (2002) 469–486.

[18] Sarah Eisenstat, Dana Angluin, Learning random DFAs with membership queries: the goodsplit algorithm, in: ZULU Workshop Organised During ICGI, 2010.

[19] M.D. Ernst, J. Cockrell, W.G. Griswold, D. Notkin, Dynamically discovering likely program invariants to support program evolution, IEEE Trans. Software Eng. 27 (2)

(2001) 99–123, A previous version appeared in ICSE99, in: Proceedings of the 21st International Conference on Software Engineering, Los Angeles, CA, USA, 1999.

[20] A. Groce, D. Peled, M. Yannakakis, Adaptive model checking, Bull. IGPL 14 (5) (2006) 729–744.

[21] Roland Groz, Muhammad-Naeem Irfan, Catherine Oriat, Algorithmic improvements on regular inference of software models and perspectives for security testing, in: ISoLA, vol. 1, 2012, pp. 444–457.

[22] Falk Howar, Bernhard Steffen, Maik Merten, From ZULU to RERS—lessons learned in the ZULU challenge, in: ISoLA, vol. 1, 2010, pp. 687–704.

[23] Hardi Hungar, Oliver Niese, Bernhard Steffen, Domain-specific optimization in automata learning, in: CAV, 2003, pp. 315–327.

[24] Muhammad Naeem Irfan, Catherine Oriat, Roland Groz, Angluin style finite state machine inference with non-optimal counterexamples, in: Proceedings of the First International Workshop on Model Inference in Testing, MIIT'10, ACM, 2010, pp. 11–19.

[25] Muhammad Naeem Irfan, Roland Groz, Catherine Oriat, Improving model inference of black box components having large input test set, in: Proceedings of the 11th International Conference on Grammatical Inference, ICGI 2012, September 2012, pp. 133–138.

[26] JabRef. Reference Manager. <http://jabref.sourceforge.net/>.

[27] jEdit. jEdit—Programmer's Text Editor. <http://www.jedit.org/>.

[28] K. Li, R. Groz, M. Shahbaz, Integration testing of components guided by incremental state machine learning, in: Testing: Academia and Industry Conference—Practice And Research Techniques (TAIC PART 2006), 2006, pp. 59–70.

[29] D. Lorenzoli, L. Mariani, M. Pezz F, Automatic generation of software behavioral models, in: 30th International Conference on Software Engineering (ICSE 2008), 2008, pp. 501–510.

[30] Erkki Mäkinen Tarja Systä, MAS—an interactive synthesizer to support behavioral modeling in UML, in: ICSE, 2001, pp. 15–24.

[31] Oded Maler, Amir Pnueli, On the learnability of infinitary regular sets, Inf. Comput. 118 (2) (1995) 316–326.

[32] Karl Meinke, Fei Niu, A learning-based approach to unit testing of numerical software, in: ICTSS, 2010, pp. 221–235.

[33] Henry Muccini, Andrea Polini, Fabiano Ricci, Antonia Bertolino, Monitoring architectural properties in dynamic component-based systems, in: CBSE, 2007, pp. 124–139.

[34] Oliver Niese, An Integrated Approach to Testing Complex Systems, PhD Thesis, University of Dortmund, 2003.

[35] Doron Peled, Moshe Y. Vardi, Mihalis Yannakakis, Black box checking, in: FORTE, 1999, pp. 225–240.

[36] Harald Raffelt, Maik Merten, Bernhard Steffen, Tiziana Margaria, Dynamic testing via automata learning, STTT 11 (4) (2009) 307–324.

[37] Ronald L. Rivest, Robert E, Schapire, Inference of finite automata using homing sequences, in: Machine Learning: From Theory to Applications, 1993, pp. 51–73.

[38] Muzammil Shahbaz, Reverse Engineering Enhanced State Models of Black Box Components to Support Integration Testing, PhD Thesis, Grenoble Institute of Technology, 2008.

[39] Muzammil Shahbaz, Roland Groz, Inferring mealy machines, in: FM, 2009, pp. 207–222.

[40] Guoqiang Shu, David Lee, Testing security properties of protocol implementations—a machine learning based approach, in: ICDCS'07: Proceedings of the 27th International Conference on Distributed Computing Systems, IEEE Computer Society, Washington, DC, USA, 2007, p. 25.

[41] B.A. Trakhtenbrot, Ya. M. Barzdin, Finite Automata, Behaviour and Synthesis, North-Holland, 1973.

[42] N. Walkinshaw, K. Bogdanov, S. Ali, M. Holcombe, Automated discovery of state transitions and their functions in source code, Softw. Test. Verif. Reliab. 18 (2008) 99–121.

[43] T. Yokomori, Learning non-deterministic finite automata from queries and counter-examples, Mach. Intell. 13, 1994, pp. 169–189.

ABOUT THE AUTHORS

Muhammad Naeem Irfan received his master's degree in software engineering from "Université de Franche-Comté" and his doctorate from "Université de Grenoble". His research interests include software model inference, optimization and analysis of algorithms, security testing, model based testing, black box testing, fuzzing and model checking.

Catherine Oriat obtained her engineering degree in computer science from "Ecole Nationale Supérieure en Informatique et Mathématiques Appliquées de Grenoble" in 1991, and doctorate in computer science from INPG, France in 1996. She is now assistant professor at Grenoble INP-Ensimag. Her research interests include testing and finite state machine inference.

Roland Groz graduated in 1980 from Ecole Polytechnique. He holds a PhD in Computer Science from Université de Rennes (on protocol verification) and HDR from U. Bordeaux (on software engineering for telecoms). From 1982 to 2002 he worked at France Telecom R&D in Lannion France. From 1997 to 2002 he was head of a research department, dedicated to software specification, prototyping and validation techniques. He was also in charge of coordinating all research in software engineering at France Telecom. In September 2002, he joined INPG (Grenoble Institute of Technology) as a professor.

His current research focuses on two main themes:

− Security of information systems and networks;

− Software components: assessment, formal modelling, partial characterizations and integration testing.

CHAPTER FOUR

Testing of Configurable Systems

Xiao Qu
Industrial Software Systems, ABB Corporate Research, Raleigh NC 27606, USA

Contents

Abstract

Configurable software system allows users to customize its applications in various ways, and is becoming increasingly prevalent. Testing configurable software requires extra effort over testing traditional software because there is evidence that running the same test case under different configurations may detect different faults. Differentiating test cases and configurations as two independent factors for testing, we must consider not just which test case to utilize, but also which configurations to test. Ideally, an exhaustive testing approach would combine every test case with every possible configuration. But since the full configuration space of most software systems is huge, it is infeasible to test all possible configurations with all test cases. Instead, selection techniques are necessary to select configurations for testing a software system, and to select test cases for the different configurations under test.

Despite successful selection techniques, sometimes it is still costly to run only selected configurations and test cases. In particular, the cost is magnified when new features and functionality are added as a system evolves, and the new version is regression tested. *Regression testing* is an important but expensive way to build confidence that software changes do not introduce new faults as the software evolves, and many efforts have been made to improve its performance given limited resources. *Test case prioritization* has been extensively researched to determine which test cases should be run first, but has rarely been considered for configurations. In this chapter we introduce issues relevant to testing configurable software systems, we then present techniques for both selection and prioritization of these systems.

1. INTRODUCTION

User configurable software—software that can be customized by a user through a set of *options* (also called *configurable options*)—is becoming increasingly prevalent. For example, Mozilla Firefox, one of the most popular web browsers, has many options that can be easily controlled by the user through a graphical user interface (GUI). A small subset of the Firefox options is shown in the window box of Options in Fig. 1. One of the options, "Block pop-up windows," enables the user to specify if pop-ups are allowed when he or she browses web pages, and the web pages will show in different formats according to different *choices*. All options of the system[1] combined compose a *configuration*, associated with respective choices selected for each option. Configurations exist in various forms so that they are controlled in different ways. Some software lets users control configurations with GUIs (for example, web browsers as just illustrated) by clicking buttons, or checking/unchecking option boxes. Some other software (for example, vim, an extended version of the vi editor) sets initial configurations by reading a resource file when the software is launched, and users can change it by issuing commands while operating on it.

This property of configurability makes software flexible and convenient for users. They can set different configurations for the system according to different requirements or preferences. However, configurable software requires extra effort in testing, because with different configurations, the same user operation may result in different software behaviors. For example, a conflict between IE7 and the Google toolbar was reported by many users [15], that they lost the Open In New Tab option in the right-click menu, when the Google toolbar was enabled. To resolve this problem, the users must disable the Google toolbar and in some cases may need to uninstall it completely. From a testing perspective, given a test case that opens a new

---

[1]In this chapter, the terms "software" or "system" are both used to refer to a software system.

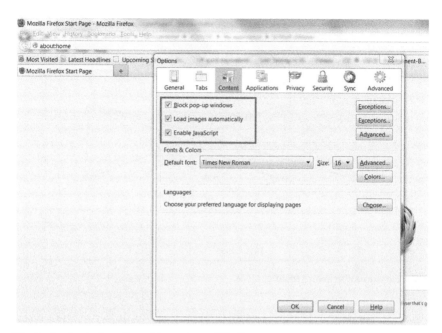

Fig. 1. A subset of configurable options of Firefox.

web page from a web link using the Open In New Tab option in the right-click menu, if we run it with the Google toolbar disabled when we set up the configurations we would not detect this incompatibility (a system fault), but if we run it with the Google toolbar enabled we would detect it.

More generally, given the same test case, running it on different configurations may expose different faults [9, 20, 26]. This implies that we need to not only consider test cases but also configurations when testing configurable systems. We call this *configuration aware testing*, in which the configuration and test case are two independent layers or factors, though they may have underlying interactions. This is different from traditional black box testing in that we select different options (parameters) for program inputs, which are regarded as part of the test cases. We term the collection of all possible configurations of a software system the *configuration definition layer* (CDL) for that system. The CDL sits on top of the normal set of inputs (test cases) to the system and therefore magnifies by a multiplicative factor (i.e., the number of configurations under test) the already large set of test cases needed for testing.

An illustration of configuration aware testing is shown in Fig. 2, which contains *m* configurations and *n* test cases under test [20]. To achieve the best testing effect, ideally we need to run each test case under each configuration

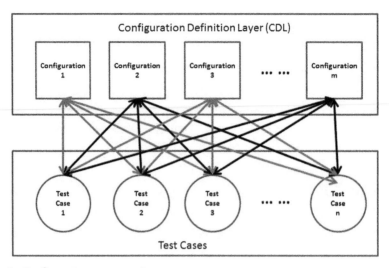

Fig. 2. Configuration aware testing.

as shown with the bold arrows in Fig. 2. Assume we want to run n test cases on m configurations, and that each test case incurs the same cost, t, regardless of the configuration, then the total cost of testing is $(t \times n) \times m$. However, this exhaustive testing is infeasible in practice due to time and resource limitations. In *bash* for example, which is a command language interpreter for the GNU operating systems, a partially modeled configuration space leads to approximately 7.6×10^{23} configurations [20]. The exponential growth of configurations with the increase in numbers of options makes it infeasible to test all possible configurations.

Alternatively, a sampling or selection technique can be used to select certain configurations that somehow "cover" an effective subset of the configurations in the CDL. As a result, the testing cost can be reduced by decreased number of configurations to test (i.e., m). Many methods are possibly applicable for selecting configurations: to create a single default configuration, to randomly select some configurations from the CDL, or to generate configurations by some systematic method where we can quantify what has been covered and tested. Among the systematic methods used to date is combinatorial interaction testing (CIT), which was originally created to sample a program's input space [6], to ensure that all t-way $(t > 1)$ combinations of choices belonging to each input parameter (option) are tested together at least once. Recent work also suggests that CIT can provide an effective way to sample configurations [10, 26], yielding higher code coverage and fault detection compared to randomly selected configurations.

Given selected configurations, for simplicity, some approaches just run the same full test suite under each selected configuration [20], but this causes redundant testing, as some test cases may behave the same under different configurations and running them repeatedly under every single configuration lead to no increase in the code coverage and fault detection. This indicates that the testing cost can be further reduced by cutting down the number of test cases (i.e., n). Techniques have been proposed to select test cases for different configurations under test [19, 21], where static analysis [2, 3] has been applied to identify the code impacted by changes between different configurations, and only test cases that cover these impacted code will be selected for testing a different configuration.

When testing configurable software, we must also consider its lifetime behavior, not just a single version. As systems evolve, before new versions are released, software must be re-tested to ensure quality, a process that is called *regression testing*. Regression testing is an expensive part of the software maintenance process, which is performed each time a system is modified and is often resource limited [12, 17]. Due to the impact of configurations on testing as described, the cost of regression testing is further magnified by the number of configurations. In other words, under some circumstances, given limited resources or time, even if selection techniques are applied, it is still possible to run out of time, therefore configurations that have higher probabilities of detecting more (important) faults, should be ordered (*prioritized*) to be tested earlier. There are different prioritization techniques for ordering configurations [20].

The rest of this chapter is organized as follows. Section 2 introduces the challenges of testing a configurable system. Section 3 identifies particular issues of testing a configurable system in a single version and across multiple versions, and summarizes common issues to be addressed by different techniques. Sections 4 and 5 introduce techniques of selecting configurations and test cases for testing a configurable system. Section 6 introduces the techniques to prioritize selected configurations in different ways. Finally, Section 7 concludes the testing of a configurable system and introduces future work.

2. CHALLENGES OF TESTING CONFIGURABLE SYSTEMS

Testing non-configurable systems usually involve challenges such as test case generation [16], test case selection [22], and test case prioritization [24].

But to test configurable systems, additional issues need to be considered at the CDL layer, due to the impact of configuration on testing effectiveness. This section first illustrates the impact and analyzes its cause.

At the code level, different configurations may control different executions of code by various methods, for example, by introducing macro definitions for configurable options. Setting different configurations (i.e., selecting different option values) turn these macros on or off (e.g., using "#ifdef, #else, and #endif"). For example, *vim*, a popular text editor, reads a *.vimrc* file to initialize its configuration. Users can make dynamic changes to the configuration by issuing a "`set` x" command on the command-line mode, where x is any one of the pre-defined configurable options. In response, the system may change its behavior. For example, there is a configurable option `updatecount` that sets the number of synchronized characters between the swap file and the disk. When `updatecount` is set zero, no swap file will be created at all. At the code level, the value of `updatecount` is passed to a macro parameter *p_uc*. There is another related option, `swapfile`, that turns a swap file on or off. At the code level, the value of `swapfile` is passed to a macro parameter *b_may_swap* that decides whether or not a swap file can be opened. These two macro parameters, *p_uc* and *b_may_swap*, control a function named *mf_sync()*, located in the source file *memfile.c*. If a swap file is not permitted by `set updatecount=0` and `set noswapfile`, then *mf_sync()* will never be executed, regardless of which test cases are executed— the given test cases simulate how users edit files (such as creating a new file or modifying an existing file in multiple ways). Suppose there is a fault located in *mf_sync()*. With the options `set updatecount=0` and `set noswapfile`, this fault cannot be detected. This problem cannot be solved by changing to different test cases; instead, the configurable options must be changed to `set updatecount>0` and `set swapfile`.

As a result of the configuration impact, it is necessary to consider both test cases and configurations when testing configurable systems. Challenges at the CDL layer may involve configuration generation, selection, and prioritization. These issues are discussed in following sections.

3. STRATEGIES IN TESTING CONFIGURABLE SYSTEMS

This section provides an overview of the concrete issues and the various strategies for testing configurable systems. It first discusses the problems and solutions of testing a single version of a configurable system, followed by a discussion of the problems and solutions in regression testing, when a system evolves. It then summarizes the strategies that are applicable in both testing environments. These problems are related to two big questions:"*what* configurations and/or test cases should be executed?" and "in which *order* the configurations and/or test cases should be executed?" (same as traditional testing[2] [24], the *order* issue is addressed only for regression testing).

This section only discusses configuration-related testing problems and techniques. Traditional test case generation, selection, and prioritization is outside the scope of this work.

3.1 Testing a Single Version

Recent studies have shown that the same test suite running under different configurations may detect different faults [9, 20]. Hence, in addition to creating effective test suites, it is also necessary to determine which configurations, from the large space of configurations, should be tested.

Ideally, all possible settings of the CDL, i.e., all possible configurations, should be tested with each applicable test case. "All possible configurations," thus, is naturally considered as the first possible choice to address the problem of "*what* **C**onfigurations are to be tested," which we denote as $What_{C1}$. However, this solution is usually infeasible in practice.

Alternatively, a subset of configurations can be selected to somehow "cover" an effective subset of configurations in the CDL. We call the selected configurations a configuration sample, and denote this solution as $What_{C2}$ for the "*what* configuration" problem. Many methods can be utilized to sample configurations. For example, some testers select random configurations from the full configuration space, while experienced testers create more practical configurations. A more systematic method called combinatorial interaction testing (CIT) [9, 10] is also effective, with the additional benefit of generating configurations automatically.

After determining what configurations are to be tested, we then determine what test cases are executed under each selected configuration. Some approaches run the same full test suite[3] for each configuration selected [20]. This is the first solution to the problem of "*what* **T**est cases are to be executed," which we denote as $What_{T1}$. However, this approach is redundant because some test cases may not be impacted or controlled by certain configurations—they behave the same, cover the same code, and have

---

[2] Traditional testing does not consider what configurations are to be tested.
[3] The generation of the initial test suite is outside the scope of this work.

the same fault detection ability when executed under these different configurations. A test case selection approach [19, 21] is proposed to select test cases for new configurations, based on a static change impact analysis [3] of configuration changes. These test cases may cover different code and have chances to detect new faults when executed under the different configurations. We denote this approach as $What_{T2}$, another solution to the "*what* test case" problem.

In summary, the issues to be addressed when testing a single version of configurable system are "what" configurations and "what" test cases need to be tested. The possible solutions are:

- $What_{C1}$: All possible configurations.
- $What_{C2}$: A subset of configurations (a configuration sample).
- $What_{T1}$: Full test suite.
- $What_{T2}$: Test cases selected based on configuration changes.

3.2 Regression Testing

When testing configurable software, we must also consider its lifetime behavior, not just a single version. As systems evolve, before new versions are released, software must be re-tested to ensure quality, a process that is called *regression testing*.

In regression testing of configurable software, we first determine what configurations and test cases are to be executed for the new version. We can use the same methods ($What_{C1}$ and $What_{C2}$) for selecting configurations[4] and the same methods ($What_{T1}$ and $What_{T2}$) for selecting test cases as we have introduced in Section 3.1.

Due to the impact of configurations on testing as described previously, the cost of regression testing increases with the number of configurations. For example, in one case study [20], each configuration in a target system required eight hours to regression test. Even if the full configuration space has been dramatically reduced from 2^{90} to 60 by the CIT sampling technique, to regression test these 60 configurations required almost three weeks. When testing a new version of software, more resources and time is usually allocated for testing new configurations and functionalities, compared to what is allocated for testing configurations from the previous version. Under this circumstance, in which resources and time is very limited for regression testing previous configurations, even if selection techniques are applied, it is still

---

[4] Testing of new configurations and features involves configuration augmentation, which is outside the scope of our discussion.

possible to run out of time. Therefore, configurations that have higher probabilities of detecting more (important) faults, should be ordered (*prioritized*) to be tested earlier.

As a result, besides the "what" question previously discussed, another issue needs to be considered that in what order are run configurations and test cases. Many prioritization techniques use historical testing data to guide prioritization for testing subsequent versions of a program. Frequently used criteria include *statement coverage, function coverage* (both are a type of *code coverage*) and *fault finding exposure*, among others [24]. Sometimes, when historical testing data is not available, prioritization can be done based on other useful and important information, such as a specification-based prioritization technique [20], in which a specification defines configurable *options* and their associated *values*. The basic idea of this technique is to test values with more complexity earlier, under the assumption that more complex features trigger more code to be executed (higher code coverage), and higher code coverage implies a higher probability of detecting more faults. These possible orders for running configurations are listed below as $Order_{C1}$ to $Order_{C4}$. The order of test cases that have been well studied [24] is outside the scope of this work:

- $Order_{C1}$: Random order.
- $Order_{C2}$: Code coverage based-reordering.
- $Order_{C3}$: Fault detection based-reordering.
- $Order_{C4}$: Specification based-reordering.

3.3 Summary of Techniques

Table 1 shows an overview of the challenges and applicable approaches (the state of the art) for testing configurable systems, at the different layers (i.e., configuration and test case), and in different testing environments (i.e., testing a single version and regression testing).

Running the same full test suite (i.e., $What_{T1}$) under all possible configurations (i.e., $What_{C1}$) in a random order (i.e., $Order_{C1}$) is simple, so that it is not described in detail in this chapter. Other solutions and techniques are

Table 1 Challenges and approaches for testing configurable systems.

Challenges		Single Version Testing	Regression Testing
Configuration Layer	Selection	$What_{C1}, What_{C2}$	$What_{C1}, What_{C2}$
	Prioritization	NA	$Order_{C1}$ to $Order_{C4}$
Test Case Layer	Selection	$What_{T1}, What_{T2}$	outside the scope
	Prioritization	NA	outside the scope

introduced in the following order: Section 4 introduces the CIT approach that generates configuration samples, referring to $What_{C2}$; Section 5 introduces the test case selection approach for different configurations, referring to $What_{T2}$; and Section 6 introduces various approaches to prioritize selected configurations, referring to $Order_{C2}$, $Order_{C3}$, and $Order_{C4}$:

- $What_{C2}$: Configuration selection (Section 4).
- $What_{T2}$: Test case selection based on configuration changes (Section 5).
- $Order_{C2}$: Configuration prioritization based on code coverage (Section 6).
- $Order_{C3}$: Configuration prioritization based on fault detection (Section 6).
- $Order_{C4}$: Configuration prioritization based on specification (Section 6).

4. CONFIGURATION SELECTION

This section introduces the technique that selects configurations from the full configuration space. Section 4.1 first introduces Test Specification Language, which is used to describe configurations in a well-defined format. Section 4.2 then describes combinatorial interaction testing (CIT) technique and its applications, particularly, on configuration sampling or selection.

4.1 Test Specification Language

The Test Specification Language (TSL) [18] provides a specification-based method for defining the combinations of factors that influence program behavior that should be tested together. TSL partitions the system inputs into *parameters*. For each parameter, a set of *choices* is defined based on equivalent classes. For example, Fig. 3 (left) shows the TSL definition of a coffee

TSL for Coffee Machine

Parameters:	
Type:	Bold
	Medium
	Decaf
Sugar:	None
	Normal
	Double
Creamer:	None
	Vanilla
	Hazelnut
Size:	Small
	Medium
	Large

Equivalent definition as TSL

Parameters	Type	Sugar	Creamer	Size
Choices	Bold	None	None	Small
	Medium	Normal	Vanilla	Medium
	Decaf	Double	Hazelnut	Large

Fig. 3. TSL definition of a vending machine.

Partial TSL for Bash

Options
    cmdhist: shopt -s
                shopt -u
    functrace: set -o
                null
    IGNOREEOF: unset
                null
    HISTSIZE: 1000
                200
                default

Equivalent definition as TSL

Options	cmdhist	functrace	IGNOREEOF	HISTSIZE
Values	shopt -s	set -o	unset	1000
	shopt -u	null	null	200
	-	-	-	default

Fig. 4. TSL definition of partial *bash* configuration.

machine's input space: there are four *parameters* (a table can also be used as shown on the right), each containing three different *choices*.

TSL can also be used to define system configurations [8–10, 26], in which each parameter is equivalent to each configurable option, and each choice is equivalent to each value of an option. For example, Fig. 4 shows an example of a partial TSL definition of the configuration for the shell interpreter *bash*. The full combination of configurable options will result in 24 configurations. This is only a very small subset of all configurable options of *bash*; in real systems, the number of configurations will grow exponentially with the increase in numbers of options and values. In real testing of *bash*, a partially modeled configuration space leads to approximately 7.6×10^{23} configurations [20].

In TSL the large combinatorial space is usually reduced through two approaches. The first approach sets specific choices as *single* or *error*, meaning that these are tested alone. In the example, we may set HISTSIZE: *1000* as *single* so that this particular choice (i.e., *1000* for option HISTSIZE) can show up only once in the generated configurations. The second approach adds properties to particular choices and defines *constraints* that relate other choices to these properties. For example, "HISTSIZE: *200* and IGNOREEOF: *unset* are not allowed to happen together or they do not need to be tested together," so this pair may never appear in the generated configurations. These approaches can significantly reduce the final set of combinations associated with a TSL specification. Alternatively, in practice, a frequently used approach of reducing the full combination space is to use combinatorial interaction testing (CIT) [6], as introduced in the following section.

4.2 Combinatorial Interaction Testing (*What$_{C2}$*)

CIT is a technique originally created to sample programs' input space. CIT sampling models the inputs (*factors*) for a software system and their associated *values* and combines these systematically so that all *t*-way ($t > 1$) combinations of values belonging to each factor are tested together, where *t* is called *the strength of testing*. For instance when $t = 2$, we call this pair-wise testing.

To model a software's input using CIT we first need to define the *factors* and their associated *values* in some format. One way to do this is with TSL as defined in the previous section, in which *parameters* and *choices* are interchangeable with *factors* and *values*, respectively. But CIT differs from TSL in that it provides a systematic sampling of the input space.

CIT samples are defined by mathematical objects called covering arrays. A *covering array, $CA(N; t, k, v)$*, is an $N \times k$ array on *v* symbols with the property that every $N \times t$ sub-array contains all ordered subsets from *v* symbols of size *t* *at least* once [7], where *N* stands for the size of the sampling (number of combined objects), *t* maintains the same definition as above, and *k* and *v* represent the total number of factors and values, respectively. Quite often in software testing the number of values for each factor is not the same. Therefore, we use the following expanded definition (often called a mixed level covering array) that uses a vector of *v*s for the factors: $CA(N; t, k, (v_1 v_2 \ldots v_k))$, where $v = \sum_{i=1}^{k} v_i$. We use a shorthand notation to describe these arrays with superscripts to indicate the number of factors with a particular number of values. For example, a pair-wise ($t = 2$) covering array with five factors ($k = 5$), three of which are binary ($v_1 = 2$) and two of which have four values ($v_2 = 4$), can be written as follows: $CA(N; 2, 2^3 4^2)$ (we remove the *k* since it is implicit) [4].

As defined, CIT techniques do not directly use the same methods as TSL to reduce the large combinatorial space, but rather reduce the combinatorial space by systematically testing only *t*-way combinations. It is possible, however, to combine these techniques by adding *single* test cases to the CIT test suite and to consider *constraints* if certain combinations are illegal [6].

As shown before, the full combination of configurable options (Fig. 4) will result in 24 configurations, while a pair-wise covering array contains only six configurations (an applicable pair-wise sample of configurations) as illustrated in Fig. 5 (right), which is a $CA(6; 2, 2^3 3^1)$. The shaded boxes are all two-way combinations (or pairs) with the value *shopt-u* for factor cmdhist, which are (cmdhist: *shopt-u*, functrace: *null*), (cmdhist: *shopt-u*, IGNOREEOF: *null*), (cmdhist: *shopt-u*, HISTSIZE: *default*) in the

Partial TSL for *bash*

Parameters	
cmdhist:	shopt-s
	shopt-u
functrace:	set-o
	null
IGNOREEOF:	unset
	null
HISTSIZE:	1000
	200
	default

Pair-wise CIT Sample

	cmdhist	Functrace	IGNOREEOF	HISTSIZE
1	*shopt-u*	null	null	default
2	shopt-s	null	null	1000
3	shopt-s	set-o	unset	default
4	*shopt-u*	set-o	null	200
5	shopt-s	null	unset	200
6	*shopt-u*	set-o	unset	1000

Shaded boxes show all 7 pairs with *shopt-u*

Fig. 5. *Bash* configurations.

first configuration, (cmdhist: *shopt-u*, functrace: *set-o*) in the fourth configuration, and so on. Previously defined constraints may also be considered so that "HISTSIZE: *200* and IGNOREEOF: *unset* are not allowed to happen together". Then the HISTSIZE: *200* or IGNOREEOF: *unset* choice of the fifth configuration should be replaced by another value, and another configuration(s) should be added to cover other missing pairs caused by this replacement.

CIT samples can be generated by various tools, associated with different strategies, such as greedy or meta-heuristic search algorithms [7, 11, 6].

Results in [9, 10] have shown that combinatorial interaction testing is effective in sampling configurations for testing, based on empirical studies on real configurable software systems, measured by fault finding ability. Specifically, Yilmaz et al. [26] apply combinatorial interaction testing for sampling configuration spaces, in order to more efficiently detect failures that are due to configuration dependencies and characterize faults, compared to exhaustively testing each configuration [14]. Cohen et al. propose an improved CIT technique for sampling configurations [8], by considering the practical concerns of constraints between combinations of options. They examine two highly configurable software systems to quantify the variety and type of constraints that can arise, and present a general constraint representation and solving technique that can be integrated with existing combinatorial interaction testing algorithms.

As far as we know, to create configurations for testing, people in industry also apply pair-wise CIT, while they also have test engineers with expert knowledge to create default configurations [21].

5. TEST CASE SELECTION (*WHAT$_{T2}$*)

Section 4 introduces the technique to select configurations for testing configurable systems. Given these selected configurations, it is redundant to run the same full test suite under each of them. But to date, there is only a small body of research addressing test case selection for different configurations.

Robinson and White [21] proposed an approach for selecting test cases based on configuration changes. A tool, Firewall, is implemented to analyze the impact of configuration changes for selecting test cases. This approach only identifies local impact which is one calling level away from the change. But in fact configuration impact usually propagates globally beyond one calling level, across different functions, files, and modules [1]. Therefore, this test case selection based on local impact may not be safe [23]. Our recent work [19] addressed the limitation of the Firewall, and also proposed a new approach to select test cases for different configurations. This section describes our approach in detail.

Our approach assumes that a test suite already exists for the system under test and that there are no changes made to the system source code when testing with different configurations. To formalize, let an existing test suite T be used for testing a system S under a configuration C, the test case selection approach decides for each $t \in T$, if t has to be used for testing S under a different configuration C'.

The essential goal of testing S under a new configuration C' is to test the code in S that has not been covered when S is tested under the old configuration C. To achieve this goal, a test case should be selected if it covers different code when executed under C' that has not been covered by the same test case when executed under C. Formally, t_i should be selected if $\text{cov}(C',t_i) - \text{cov}(C,t_i) \neq \emptyset$, where $\text{cov}(C,t_i)$ denotes the code covered by running test t_i on configuration C. However, the dynamic $\text{cov}(C',t_i)$ is unavailable in practice unless we re-run the full test suite T under C'. Given that t_i and the source code of S are unchanged, the different coverage of t_i in S under C' can only be caused by the configuration changes. Hence, our approach conservatively approximates the dynamic coverage difference (i.e., $\text{cov}(C',t_i) - \text{cov}(C,t_i)$), which is not available in practice, using the static impact of configuration changes. Accordingly, our approach selects t_i for testing S under configuration C' if t_i covers, under configuration C, parts of the source code that are impacted by the configuration changes (denoted as Δ). Formally, t_i is selected if $\text{cov}(C,t_i) \bigcap \text{imp}(\Delta) \neq \emptyset$.

Code coverage can be measured at different granularities, such as statement, block, or function. Our approach uses function-level granularity because a function is typically considered as a unit in testing. In summary, our approach selects t_i for testing S under C' if

$$\mathrm{cov}^f(C, t_i) \cap \mathrm{imp}^f(\Delta) \neq \emptyset. \tag{1}$$

Our approach involves four steps as shown in Fig. 6: (1) computing configuration differences between C and C', (2) computing the functions impacted in S by the configuration differences, (3) computing the function coverage for each test case t_i in T when S is tested under C dynamically, and (4) selecting t_i based on Eqn 1.

Figure 7 defines a configurable system S under test. S has eight functions from $f1$ to $f8$, and it takes an integer x as its input. The control flow graph (CFG) is shown in the left part of Fig. 7. S contains three configurable options, $P_1, P_2,$ and $P_3,$ each of which has two values, ON and OFF (as shown in the right top table in Fig. 7). The test suite for testing S contains three test cases, $t1, t2,$ and $t3$ (as shown in the right bottom table in Fig. 7). Suppose we have tested S by executing the full test suite under a configuration C, in which value ON is assigned to $P1$, while value OFF is assigned to $P2$ and $P3$ (as shown in the right middle table in Fig. 7). Given a new configuration C', in which ON is assigned to $P2$ while OFF is assigned to other two options, we illustrate how test cases are selected for testing S under C' using the approach:

1. *Computing Code Coverage:* While executing T under C, our approach computes $\mathrm{cov}^f(C, t_i)$ for each $t_i \in T$. Table 2 shows the resulting function coverage matrix. Value 1 (0) represents that a function is covered (not covered) by a test case under a specific configuration. In our example, $\mathrm{cov}^f(C, t1) = \{f1, f3\}, \mathrm{cov}^f(C, t2) = \{f5\},$ and $\mathrm{cov}^f(C, t3) = \{f7\}$.
2. *Computing Configuration Differences (Δ):* For each option $P_i(i \in [1, m], m$ denotes the number of configurable options), our approach compares its

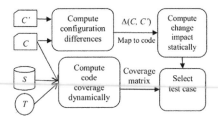

Fig. 6. Overview of test case selection approach.

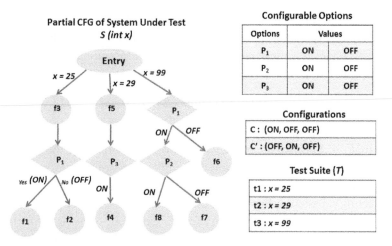

Fig. 7. An example configurable system.

Table 2 Function coverage under configuration C.

	f1	f2	f3	f4	f5	f6	f7	f8
t1	1	0	1	0	0	0	0	0
t2	0	0	0	0	1	0	0	0
t3	0	0	0	0	0	0	1	0

values in C and C': if they are different, P_i is included in set Δ. In our example, $m = 3$ and $\Delta = \{P_1, P_2\}$.

3. *Computing Configuration Change Impact* $(imp^f(\Delta))$: For each $P_i \in \Delta$, our approach computes its $imp^f(P_i)$ (implementation details are described in [19]). By combining all impact sets,

$$imp^f(\Delta) = \bigcup_{P_i \in \Delta} imp^f(P_i). \tag{2}$$

In our example, $imp^f(P1) = \{f1, f2, f6\}$ since $P1$ impacts $f1, f2$, and $f6$ by controlling their executions. Similarly, $imp^f(P2) = \{f7, f8\}$. Therefore, $imp^f(\Delta) = imp^f(P1) \bigcup imp^f(P2) = \{f1, f2, f6, f7, f8\}$.

4. *Selecting Test Cases:* According to Eqn 1, our approach selects $t1$ since $cov^f(C, t1) \cap imp^f(\Delta) = f1 \neq \emptyset$. For the same reason, $t3$ is also selected.

6. CONFIGURATION PRIORITIZATION

As introduced in Section 3.2, the impact of configurability can be particularly large in the context of *regression testing*. Just as test cases can be prioritized, so can configurations, with a goal to order the configurations in a manner that helps meet testing objectives (e.g., fault detection) earlier. But to the best of our knowledge, though a few work is loosely related to, there is little work on explicitly prioritizing configurations. For example, Mariani et al. [13] addressed regression testing in component-based systems, where each component under test may be exchangeably regarded as one single configurable option. The authors proposed techniques to automatically generate compatible and prioritized test suites based on behavioral models that represent component interactions, while executing the original test suites on previous versions of target systems. However, they did not explicitly prioritize configurations.

This section introduces our approach for explicitly prioritizing configurations, based on the *interaction benefit* [5] between configurable options, or the importance of configurable option interactions, along with several different heuristics [20]. The following descriptions starts with a introduction to the basic algorithm of interaction benefit-based prioritization.

6.1 Interaction Benefit-Based Prioritization

The interaction benefit-based prioritization approach proposed by Bryce and Colbourn [5] is a greedy algorithm for re-generating prioritized test suites which is designed to consider interaction benefit for test cases generated by a pair-wise CIT technique. The test suites generated are a special kind of a covering array called a *biased covering array*. The algorithm begins by defining a set of *weights* for each *value* of each *option* (or input parameter), reflected each value's relative importance. For each option, the weight of combining it with each other factor is computed as its total *interaction benefit (interaction weights)*. The options are sorted in decreasing order of interaction benefit and then handled as follows.

First, the individual interaction weight for each of the option's values is computed. The value of the option that has the greatest value interaction benefit is selected. After all options have been fixed and a single test is added, the interaction benefits for options are recomputed and the process starts again. The algorithm is complete when all pairs have been covered. The pseudo-code for this algorithm is presented in Algorithm 1.

The prioritization of configurations follows the same procedure [20], just by replacing the prioritized object from test case to configuration.

The key element of this approach is the weight of each value of each option, which reflects its importance. Importance can be measured in different aspects. For example, a value that yields a higher code coverage may be regarded as more important; or a value that detects more faults can be regarded as more important. Different weighting heuristics are introduced in [20]. We briefly introduce their main concepts in the following sections.

Algorithm 1. Pair-wise Re-generation Algorithm [5]

RemainingPairs=AllPairs of Options;
while (RemainingPairs){
    Compute Interaction Weights for Options;
    Order Options;
    foreach (i=1 to numOptions)
    {
      Select HighestWeightUnfixedOption;
      foreach (j=1 to numValues){
        Compute Interaction Weights for Option Values;
        Select Value with highest weight;
      }
    }
    Add Test Case (or Configuration);
}

6.2 Coverage- and Fault Detection-Based Weighting

The coverage-based weighting ($Order_{C1}$) and fault detection-based weighting ($Order_{C2}$) uses the coverage and fault detection data resulting from testing a prior version. When testing a prior version, the test cases are executed under all different configurations, hence the code coverage data (i.e., the percentage of code covered by the test cases), as well as the fault detection data (i.e., the number of faults detected by the test cases), is collected for each configuration. Since the test cases are the same for different configurations, the differences of code coverage and fault detection are caused only by the differences between configurations. In other words, given the same set of test cases, different configurations have different abilities to cover the code under test, and different capabilities to detect faults. From the perspective of regression testing, in order to achieve high code coverage and fault detection as early as possible, configurations that yield higher code coverage or fault detection are preferred to be tested earlier (we also regard them more important).

Configurations are different because different values of configurable options are selected. Hence, the values selected for the more important configuration are weighed higher. The details of computing the weights are illustrated in [20].

6.3 Specification-Based Weighting

There may be instances where we do not have prior code coverage data or fault detection data. Since we define our configuration model using TSL (Section 4.1) we can use weighting on it directly ($Order_{C3}$). In TSL-based prioritization we have the advantage of not requiring a prior version; instead we can rely on our existing version of the software to produce information to drive the prioritization.

For each parameter in the TSL specification (i.e., each configurable option of the system), we examine that parameter's possible choices (i.e., values of each configurable option). In the case of binary choices, where one choice turns a feature on and one turns a feature off, we set the choice *on* to a higher weight (for example, 0.9) and the *off* choice to a lower weight (for example, 0.1). Our intuition is that the *on* option may cause more code to be executed—this concept is consistent with code coverage-based heuristic (Section 6.2).

In cases where we have multiple choices for a parameter, we use a greater number of features or higher complexity of the choice as a proxy for higher code coverage. For instance in the system *vim* that is a text editor, there is a parameter called `laststatus` which determines when the last window will contain a status line. There are three choices: *never, only when 2 windows*, and *always*. We assign the highest importance (for example, 0.5) to the last choice, a medium importance to the second (for example, 0.3), and the smallest importance (for example, 0.1) to the first. Our intuition is that to determine and illustrate a status line is a functionality implemented by some code—if a configuration *never* leads tests to call this functionality, this code is not executed, hence this *never* choice is set the lowest weight due to its potential lower code coverage; if a configuration *always* requires tests to execute this functionality, it is potentially results in a higher coverage than other choices, hence this *always* choice is set the highest weight.

7. CONCLUSION

Testing configurable systems require extra effort over testing traditional systems because we need to not only consider test cases, but also possible configurations of the system. This chapter explored various techniques to address the challenges of testing configurable systems in different environments. It first analyzed the challenges of testing configurable systems due to the impact of configurations. After identifying various problems in testing a single version of a system, and problems with regression testing a system when it evolves, it then described current approaches that address these problems.

Although a large body of research work has started to address the challenges of testing configurable systems, there are still many problems that have to be investigated. For example, the selection and prioritization techniques introduced in this chapter assume that each test case takes the same amount of time to run for each configuration, where the effectiveness is measured by the reduction in numbers of configurations and/or test cases. However, in practice, the same test case may require a different amount of time to run for each configuration. Furthermore, the execution time of each test cases may vary considerably under the same configuration. To perform effective and efficient configuration aware testing, in terms of practical metrics such as actual testing time, cost-aware techniques are needed.

Prior studies have been conducted on testing individual software systems, but additional challenges in configurable system testing exist. For example, software developed as one product instance in a product line is usually composed of various components, where each component may be configured with multiple options. Due to complex interactions among components, prior testing approaches may be inadequate to cover all user requirements. Hence, to test this type of software involves additional challenges.

Finally, complex ultra–large-scale systems (ULSS) are becoming prevalent in many industries, such as in the smart grid [25]. These systems consist of multiple large configurable systems working together as a single entity. Existing approaches of testing single configurable systems may suffer from scalability or redundancy problems. As product lines and ULSS become a more and more popular strategy for software operations, effective and efficient testing techniques will be essential.

REFERENCES

[1] M. Acharya, B. Robinson, Practical change impact analysis based on static program slicing for industrial software systems, in: Proceedings of the International Conference on Software Engineering, Software Engineering in Practice Track, May 2011, pp. 746–765.
[2] L.O. Andersen, Program analysis and specialization for the c programming language, Technical Report, 1994.
[3] R.S. Arnold, Software Change Impact Analysis, IEEE Computer Society Press, Los Alamitos, CA, USA, 1996.

[4] R. Brownlie, J. Prowse, M.S. Phadke, Robust testing of AT&T PMX/StarMAIL using OATS, AT& T Tech. J. 71 (3) (1992) 41–47.

[5] R. Bryce, C. Colbourn, Prioritized interaction testing for pair-wise coverage with seeding and constraints, J. Inform. Softw. Tech. 48 (10) (2006) 960–970.

[6] D.M. Cohen, S.R. Dalal, M.L. Fredman, G.C. Patton, The AETG system: an approach to testing based on combinatorial design, IEEE Trans. Softw. Eng. 23 (7) (1997) 437–444.

[7] M.B. Cohen, C.J. Colbourn, P.B. Gibbons, W.B. Mugridge, Constructing test suites for interaction testing, in: Proceedings of the International Conference on Software Engineering, May 2003, pp. 38–48.

[8] M.B. Cohen, M.B. Dwyer, J. Shi, Constructing interaction test suites for highly-configurable systems in the presence of constraints: a greedy approach, IEEE Trans. Softw. Eng. 34 (5) (2008) 633–650.

[9] M.B. Cohen, J. Snyder, G. Rothermel, Testing across configurations: implications for combinatorial testing, in: Proceedings of the International Workshop on Advances in Model-based Testing, November 2006, pp. 1–9.

[10] D. Kuhn, D.R. Wallace, A.M. Gallo, Software fault interactions and implications for software testing, IEEE Trans. Softw. Eng. 30 (6) (2004) 418–421.

[11] Y. Lei, R. Kacker, D.R. Kuhn, V. Okun, J. Lawrence, IPOG: a general strategy for t-way software testing, in: IEEE International Conference and Workshop on the Engineering of Computer-Based Systems, 2007, pp. 549–556.

[12] H. Leung, L. White, Insights into regression testing, in: Proceedings of the International Conference on Software Maintenance, October 1989, pp. 60–69.

[13] L. Mariani, S. Papagiannakis, M. Pezze, Compatibility and regression testing of COTS-component-based software, in: Proceedings of the International Conference on Software Engineering, 2007, pp. 85–95.

[14] A. Memon, A. Porter, C. Yilmaz, A. Nagarajan, D. Schmidt, B. Natarajan, Skoll: distributed continuous quality assurance, in: Proceedings of the International Conference on Software Engineering, May 2004, pp. 459–468.

[15] B. Moolenaar, Lavasoft Support Forums. <http://www.lavasoftsupport.com>.

[16] M. Prasanna, S. Sivanandam, R. Venkatesan, R. Sundarrajan4, A survey on automatic test case generation, Acad. Open Internet J. 15 (2005).

[17] K. Onoma, W.-T. Tsai, M. Poonawala, H. Suganuma, Regression testing in an industrial environment, Commun. ACM 41 (5) (1988) 81–86.

[18] T.J. Ostrand, M.J. Balcer, The category-partition method for specifying and generating functional tests, Commun. ACM 31 (1988) 678–686.

[19] X. Qu, M. Acharya, B. Robinson, Impact analysis of configuration changes for test case selection, in: Proceedings of the International Symposium on Software Reliability Engineering, 2011, pp. 140–149.

[20] X. Qu, M.B. Cohen, G. Rothermel, Configuration-aware regression testing: an empirical study of sampling and prioritization, in: Proceedings of the International Symposium On Software Testing and Analysis, 2008, pp. 75–86.

[21] B. Robinson, L. White, Testing of user-configurable software systems using firewalls, in: Proceedings of the International Symposium on Software Reliability Engineering, 2008, pp. 177–186.

[22] G. Rothermel, M.J. Harrold, Analyzing regression test selection techniques, IEEE Trans. Softw. Eng. 22 (8) (1996) 529–551.

[23] G. Rothermel, M.J. Harrold, A safe, efficient regression test selection technique, ACM Trans. Softw. Eng. Methods 6 (2) (1997) 173–210.

[24] G. Rothermel, R. Untch, C. Chu, M.J. Harrold, Prioritizing test cases for regression testing, IEEE Trans. Software Eng. 27 (10) (2001) 929–948.

[25] Wikipedia, Smart Grid, 2009. <http://en.wikipedia.org/wiki/Smart_grid>.

[26] C. Yilmaz, M.B. Cohen, A. Porter, Covering arrays for efficient fault characterization in complex configuration spaces, IEEE Trans. Softw. Eng. 31 (1) (2006) 20–34.

ABOUT THE AUTHOR

Xiao Qu holds a Ph.D. with a Software Engineering specialty from University of Nebraska-Lincoln. She received her bachelor's and master's degree in Computer Science from Sichuan University, China. She has served on the program committees for several software engineering conferences and workshops, as well as reviewers for numerous software engineering conference and journal papers.

She is currently working as a Software Engineering Research Scientist in ABB Corporate Research. Her research interests include automatic software testing, especially in testing configurable software systems, applying and prioritizing combinatorial interaction testing, and search-based software engineering. She has interned at Microsoft Research.

CHAPTER FIVE

Test Cost-Effectiveness and Defect Density: A Case Study on the Android Platform

Vahid Garousi*,†, Riley Kotchorek*, and Michael Smith*

*Software Quality Engineering Research Group (SoftQual), Department of Electrical and Computer Engineering, University of Calgary, 2500 University Drive NW, Calgary, AB Canada T2N 1N4
†Graduate School of Informatics, Middle East Technical University (METU), Ankara, Turkey

Contents

Advances in Computers, Volume 89
ISSN 0065-2458, http://dx.doi.org/10.1016/B978-0-12-408094-2.00005-9

Abstract

The Android operating system is one of the most popular open-source platforms in the mobile operating system market. It had a worldwide smart-phone market share of 68% at the second quarter of 2012. However, there has been little research on test coverage and test cost-effectiveness in this platform. The goal of this case study reported in this paper is to assess test coverage, fault detection effectiveness, test cost-effectiveness, and defect density in code-base of version 2.1 of the Android platform. We raise and address five research questions (RQs) in this study. Among our results are: (1) in contrary to what one would expect, for packages with larger coverage values (meaning more rigorous testing), it is not necessarily true that less defects have been reported by the users after release. Also, it is not necessarily true that components with low coverage have more defects; (2) we re-confirm (replicate) the existence of correlation between code coverage and mutation score, similar to existing studies; and (3) the package with the highest defect density (DD) in the Android code-base is *Music* (DD = 0.19 per 1 KLOC) and the package with the lowest DD value is *ContactsProvider* (DD = 0.0003). Results of our study will help us and other researchers to get a better view on test coverage, fault detection effectiveness, test cost-effectiveness, and defect density in Android code-base.

1. INTRODUCTION

The Android operating system was originally created by Android Inc., later acquired by Google. It is an open-source project [1], maintained by the Open Handset Alliance. It is now one of the major players in the mobile operating system market. Android continues to grow its share of the mobile market with many different manufacturers [2] continually producing additional Android-based devices. According to International Data Corporation (IDC) [3], as of the second quarter of 2012, Android had a worldwide smart-phone market share of 68%, while the iOS market share was only 16.9%.

With a constantly growing user base and developer community, ensuring a high-quality software platform for Android with minimum number of defects is very important. Various studies have been conducted and published on various software quality aspects of Android, e.g., [4–12]. The goal of this paper is to conduct additional and complementary analyses in the context of test coverage, fault detection effectiveness, test cost-effectiveness, and defect density in code-base of the Android 2.1 platform (code name: Éclair).

It is anticipated that the results of the case study reported in this article will be useful to researchers and developers currently contributing or intending to contribute to the open-source Android operating system. By analyzing its test suites, developers may be able to suggest corrections for any existing defects and shortcomings; thus increasing both overall product quality and the power of its test suite. This study is also directed toward developers of other large operating system-like projects in that they can better examine how a relatively new and successful operating system is organized and can be tested.

The remainder of this paper is organized as follows. Section 2 discusses the related work. Section 3 describes the case study design and setup. An overview of the Android code-base and its automated test suites is provided in Section 4. Study results are presented in Section 5. Finally, conclusions and future work are discussed in Section 6.

2. RELATED WORKS

The current work relates to the following four topic areas:
- Test cost-effectiveness (coverage versus fault detection effectiveness of an automated test suite).
- Defect density in the Android platform.
- Mining software repositories (MSR) in the context of the Android platform.

We discuss the related works in each of the above areas next, and discuss how our work relates to them.

2.1 Test Cost-Effectiveness (Coverage Versus Fault Detection Effectiveness)

One of the popular issues (both from academic and industrial perspectives) in software testing is experimental cost-benefit analysis and fault detection effectiveness of coverage criteria and test techniques. The works reported [13–19] are among the key references in this area.

The work by Hutchins et al. [13] is one of the earliest works in this area. Using a series of experiments, the authors reported the effectiveness of data flow- and control flow-based test adequacy criteria. Another work in this context has been conducted by Andrews et al. [14], in which again mutation analysis has been used for assessing and comparing block, decision, C-use, and P-use coverage. Frankl et al. [15] compared the all-uses coverage versus mutation testing. Gupta and Jalote presented in [16] a mutation-based approach for experimentally evaluating both fault detection effectiveness and efficiency of coverage criteria. By efficiency, they meant the average testing cost for detecting a fault in a program. The results suggested that there is a trade-off between effectiveness and efficiency of a coverage criterion. Specifically, they found that the predicate coverage is the most effective at detecting faults but the least efficient whereas the block coverage was the most efficient but least effective.

Seo and Choi [18] compared five selected BB testing methods: use-case driven testing, BB testing using collaboration diagram, testing using extended use-cases, and testing using formal specifications (OCL and Object-Z), on two target SUTs. A recent industrial case study in our research group [17] is another work in this topic, in which an automated unit test framework for an industrial SCADA software was developed. The cost effectiveness of the approach and also the test suite was evaluated. Authors of [19] reported another empirical study on the fault detection ability of two WB metrics: branch and definition-usage. We conduct cross-comparison of our results with several of the above studies [13–19] in Section 5.4 whenever possible (i.e., when the data are comparable).

In our work, we will measure code coverage (Section 5.1) by running Android test suites, will measure (in Section 5.2) the fault detection effectiveness of the test suites in finding artificially-injected defects, and will mine the Android bug repository for the number of actual real defects for each package (Section 5.3). Based on these measurements, we will conduct experimental cost-benefit analysis of code coverage in Sections 5.3 and 5.4.

2.2 Defect Density

Defect density is the number of defects detected in a software component during a defined period of development/operation divided by the size of the software component [20].

The industry average for defect density is a subjective discussion and varies by the criticality of software systems. Different studies have reported values between 10 and 50 defects per 1,000 lines of code (KLOC) [21]. According to [22], for software applications developed by Microsoft, defect density is about 10–20 defects per KLOC during in-house testing and 0.5/KLOC in released products.

Focusing on defect density in the Android platform, we were able to find the following studies: [4–6]. Coverity Inc. (a major player in the software testing industry) has published two recent reports on the topic in 2010 and

2011 [4, 5]. According to [4], Coverity selected (randomly) 359 oftware defects in the Android Froyo (Unix) kernel that is used in the HTC Droid Incredible smart-phone. Of those defects, Coverity has identified 88 defects or about 25 percent of the defects, as being high-risk and potentially leading to security risk for Android users. Coverity reported that, for the Android kernel, the defect density was about 0.47 per KLOC.

According to Coverity, the defect density in Android is not that high when compared to other major open-source software code-bases that they have examined, and the industry average for defect density.

The more recent report in [5] is the results of applying automated testing to the Android Unix kernel. This study reported the defect density values for different components of the kernel, which varied between 0.41 and 0.78 defects per KLOC. In total, the firm found 359 defects in the Android kernel that runs inside the phone model "HTC Droid Incredible." As a good indicator, the report found that the Android kernel defect density is 0.47 defects per KLOC, which is better than the industry average (about 10 defects per KLOC).

Maji et al. [6] conducted a case study to characterize failures in the two mobile OSs: Android and Symbian. The study was done based on 628 bug reports from Android and 153 bug reports from Symbian. The study concluded that the following applications are the most error-prone in both these systems: development tools, web browsers, and multimedia applications. The authors further analyzed 233 bug fixes for Android and categorized the different types of code modifications required for the fixes. The analysis showed that 77% of errors required minor code changes, with the largest share of these coming from modifications to attribute values and conditions. The final analysis of [6] focused on the relation between customizability, code complexity, and reliability in Android and Symbian. The authors found that despite high cyclomatic complexity, the bug densities in Android and Symbian are surprisingly low, compared to other commercial software systems.

2.3 Mining Software Repositories (MSR) for the Android Platform

Mining Software Repositories (MSR) has become a very active research area in recent years [23]. Software practitioners and researchers are recognizing the benefits of mining information in software repositories (such as source control systems, archived communications between project personnel and defect tracking systems) to support the maintenance of software systems and improve software design/reuse. Research is now proceeding to uncover the ways in which mining these repositories can help to understand software development and software evolution, to support predictions about software development, and to exploit this knowledge concretely in planning future development projects.

There are many works in the MSR literature which focus on mining test-related artifacts (e.g., [24–26]). For example, Zaidman et al. applied data mining on source control systems to study co-evolution of production and test code [25]. Barahona and Robles applied social network analysis to the information in CVS repositories to see which developers and testers collaborate more often with each other [26]. Our previous work in [24] used data mining to extract insights about issue management processes in open-source projects, e.g., how responsive the development teams are in handling bug reports.

There have been several works in the MSR literature focusing on the Android platform, e.g., [7–12]. The work in [7] investigated the bug introducing changes in Android. The study [8] intended to find the "trendy" bugs in the Android bug repository and the topic trends in the Android bug reports. The work in [9] conducted a multidimensional analysis of Android's layered architecture using the Android bug repository and the people involved in submitting and handling bugs. It reported three findings [9]: (1) while some architectural layers have a diverse interaction of people, attracting not only non-central reporters but highly important ones, other layers are mostly captivating for peripheral actors; (2) the study exposed that even the bug lifetime is similar across the architectural layers, some of them have higher bug density and differential percentages of unsolved bugs; and (3) comparing the popularity distribution between layers, the also identified one particular layer that is more important to developers and users alike.

The work in [10] mined the Android's source code repository and extracted the build dependency perspective of Android's architecture. [11] mined and reported the change history of the Android project. Finally, [12] mined for localization in Android.

3. DESIGN OF THE CASE STUDY

The design of our case study is outlined below. We discuss the goal and research questions of the study. In designing our study, we have benefited from the guidelines provided by [27–29].

3.1 Goal

This case study follows the Goal, Question, Metric (GQM) approach [30]. A research goal to explore the Android platform is presented. Research questions are devised which, when answered, will show how to attain the presented goal. In order to answer the questions at hand, various software metrics are then collected from the Android code-base, test suites, and bug repository.

Formulated using the GQM goal template [30], the goal of this case study is to assess test coverage, fault detection effectiveness, test cost-effectiveness, and defect density in code-base of the Android 2.1 platform in the context of software comprehension, testing and maintenance activities from the point of view of researchers and software developers, maintainers, and testers.

3.2 Research Questions

Based on the above goal, we raise the following research questions (RQs):

- *RQ 1:* What is amount of test coverage achieved by Android's test suites? (Section 5.1).
- *RQ 2:* How effective are the Android test suites at detecting faults? In other words, how effective are the test suites at detecting artificial faults? (Section 5.2).
- *RQ 3:* Do code coverage values correlate with actual reported number of defects? We would expect that, for packages with larger coverage values (meaning more rigorous testing), less defects are reported by the users. Do components with low coverage have more defects? Or from another perspective, do the development/testing teams tend to develop more comprehensive test suites with more coverage for components with more defects? (Section 5.3).
- *RQ 4:* As a follow-up to RQ 3: Do code coverage values correlate with rate of detecting artificial faults (mutation score)? This RQ relates to test cost-effectiveness, i.e., how does test coverage (as a notion of test cost) relates to mutation score (as a notion of effectiveness)? (Section 5.4).
- *RQ 5:* What is the defect density of different packages? How do size metrics (LOC) of packages correlate with the number of defects reported for them? (Section 5.5).

Note that RQ 1 and 2 are measurement-related. Based on the measurement data provided from RQ 1 and 2, RQs 3, 4, and 5 will conduct the necessary analysis. All the RQs are addressed in Section 5.

4. AN OVERVIEW OF THE ANDROID CODE-BASE AND ITS AUTOMATED TEST SUITES

An overview of Android's architecture is presented in Section 4.1. The size and complexity metrics of the code-base are presented in Section 4.2. We then utilize software visualization techniques in Section 4.3 to present an overview of the code-base. An overview of Android's automated test suites is presented in Section 4.4.

4.1 Architecture

The Android system is organized into a five-layer software stack [31], as shown in Fig. 1.

This software Architecture provides an abstraction of the device's hardware and its software libraries which are used by applications to interact with users. At the bottom of the stack is the Linux kernel which interacts directly with the devices and hardware [31], and has been written in C/C++. At the top of the stack sits the applications, written in Java [31], which communicate with users. The three layers in between exist to bridge the gap between the top-most layer and the bottom-most layer and to assist in

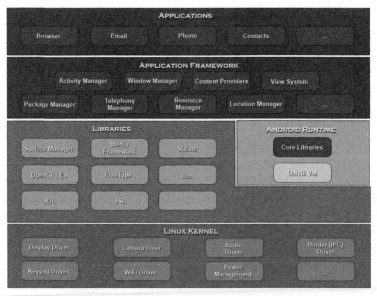

Fig. 1. Android's architecture diagram showing the five-layer software stack (taken from [31]).

the communication from the Java-based components to the native C/C++ components and vice versa.

4.2 Size and Complexity Metrics

To show the relative sizes of the various packages in the platform, we gathered the size metrics of the code. Table 1 shows the results from running the Microsoft LOC Counter tool [32] on the source code obtained from the Android Git repository. The LOC measures do not include comments in source code.

It can be seen that C and C++ make up a large portion of the code-base at about 63% of the total. Based on Android's upper software stack layers being implemented in Java, it makes sense that Java has the second highest LOC count at 22%. Android uses manifest XML files to describe different modules to the build system and the *SQLite* system as its database management system, which is why XML files have a somewhat large presence in the code-base. The web-related files are classified as any files which have the extension ASPX, ASCX, CSS, HTM, HTML, ASP, or JS. From inspecting the source code, we found that a large portion of these files are used to document various packages and the code therein.

We also aimed at gathering size and McCabe complexity metrics [33] at individual Java method levels with the intent of comparing the expected number of tests needed to fully validate the code compared to the actual number of tests provided. We used the Eclipse "Metrics" plug-in [34] and results can be seen in Table 2. A total of 71,821 Java methods were examined. Recall that the McCabe complexity of a Java method indicates the number of control flow paths inside it. The average McCabe complexity value is approximately 2.5. This means that, on average, each Java method in

Table 1 Total lines of code per language for Android 2.1 code-base.

Total LOC	9,344,603		
Number of files	54,164		
Total lines of comments	4,673,630		Percent of total (%)
LOC by language	C/C++	5,925,096	63.41
	Java	2,090,904	22.38
	SQL	124,094	1.33
	XML	632,456	6.77
	Web files	572,053	6.12
	Total	**9,344,603**	**100.00**

Table 2 Average and maximum LOC count and McCabe complexity metrics for Java methods in Android 2.1.

Average method LOC	8.78
Average method McCabe complexity	2.54
Maximum method LOC	2857
Maximum method McCabe complexity	284

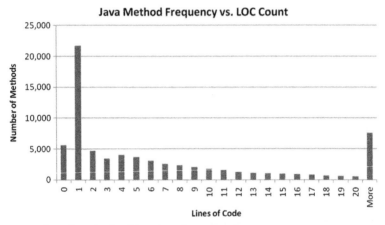

Fig. 2. Number of methods with a given lines of code count.

Android would require approximately 2.5 test cases to be fully covered [33]. Based on our count of 71,821 methods, approximately 179,553 total test methods would have to be performed in order to fully cover the methods investigated. In Section 4.4.2, the actual number of test methods is measured and is compared to this approximate required number of test methods.

Figure 2 displays a histogram of the number of methods for a given LOC count. A histogram showing the number of methods with a given McCabe cyclomatic complexity is shown in Fig. 3. As can be seen in both figures, a large portion of the Java methods inspected resides in the lower end of both spectrums. This is a positive sign as it means that most of the Java methods in the system are relatively small in terms of LOC which means they are most likely easy to understand, modify, and test.

One possible explanation for the large amount of 1-LOC methods in Fig. 2 is the existence of many object-oriented getter or setter methods. These are usually one-line long and simply set (a setter method) or get (a getter method) the value of a class attribute.

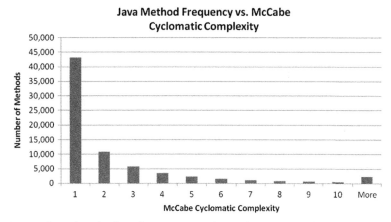

Fig. 3. Number of methods with a given McCabe Cyclomatic Complexity.

Setters usually follow the format: `void setAttributeName (Object newValue) attribute = newValue;`. Using this generally accepted naming scheme, we carried out a search to find the number of setters within the source code. The search was performed targeting any occurrence of the string "void set" in the Android code-base.

Searching for getters within the Android source proved more difficult because the return value of getter functions can be of any type. To limit the number of false positives from simply searching for `"get, "` extended regular expressions (regex) were used, e.g., we searched for `"int get"` `"double get"` and other similar patterns.

Using the above method, we were able to achieve a rough estimate of the number of accessor (getter or setter) methods within the Java-based components of Android. It was found that there are about 9,754 setters and 22,284 getters. While these values are approximate, this does suggest that most of the one-line Java methods within Android are probably accessor methods, with some accessors being more than one line.

Upon closer inspection of the methods used in calculating the values used in Fig. 2, it was found that the methods with zero lines of code occur in large part due to two reasons: (1) Empty default constructors, and (2) Native method prototypes. The native method prototypes are used to define C/C++ functions which can be called through the Java native interface and are not actual methods themselves, for example: `public native void someNativeFunction()`. A search was performed to determine approximately how many of these prototype methods were included in the collected data. This search followed the same idea from the accessor searches above. Once again, using the command-line search tools provided with Ubuntu 10.04, the strings matching the regular expression pattern "(public|private|protected) [A-Za-z0-9]* native [A-Za-z0-9_$]*[(]" were located and counted. In total, about 2,439 native prototype methods exist within the Java classes examined here.

In Fig. 2, the column labeled "more" represents all methods with greater than 20 LOC. As can be seen in the chart, there are a fair number of such methods. A lot of opportunity for refactoring was seen upon inspection of one of these methods, onTransact() from the class android. app.ActivityManagerNative, a method which handles interprocess communication with respect to Activities. The method itself is 950 Java LOC and is comprised entirely of a single switch-case statement which contains 96 different cases. To increase the maintainability and testability of this method, the work done by this method could most likely be broken down into several helper methods in order to better distribute the size and workload, especially considering there are only 7 other methods in this class at a total of 35 LOC.

Last but not least, Fig. 4 visualizes, as a scatter plot, the McCabe cyclomatic complexity of methods versus their LOC size. The number of methods (points) in this chart is 71,821. It looks visually that there is a decent correlation between McCabe complexity of methods versus their LOC size. To assess this observation quantitatively, we calculated the Pearson correlation

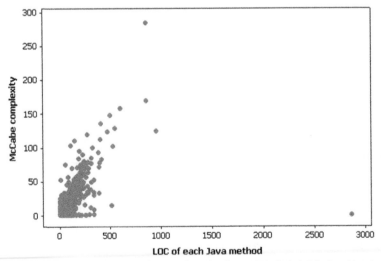

Fig. 4. McCabe cyclomatic complexity of methods versus their LOC size. Number of methods (points) = 71,821.

Table 3 The two points (methods) with the maximum values on either of the axes in the data set shown in Fig. 4.

Package	Java Class Name	Method Name ✦	LOC	McCabe Complexity
com.android. server	Window Manager Service. java	performLayout AndPlaceSurfaces LockedInner	855	284
org.ccil. cowan. tagsoup	HTMLSchema. java	HTMLSchema	2857	1

of the two data sets and the value is 0.74 (with P-value $= 0.00$), denoting the existence of a strong correlation, i.e., the larger the LOC size of a method, the larger its McCabe complexity is expected to be.

The two points (methods) with the maximum values on either of the axes in the data set are shown in Table 3. With 855 line of Java code, method `performLayoutAndPlaceSurfacesLockedInner` has a very complex code with McCabe measure of 284. On the other hand, method `HTMLSchema` is so large (LOC = 2,857) but only has one control flow path (McCabe measure = 1).

4.3 Software Visualization of the Code-Base

Software visualization is the static or animated 2-D or 3-D visual representation of information about software systems based on their structure, size, history, or behavior [35]. CodeCity [36] is an integrated environment for software analysis, in which software systems are visualized as interactive, navigable 3-D cities. The classes are represented as buildings in the city, while the packages are depicted as the districts in which the buildings reside which permits a better understanding of the relative sizes of the structures present in Android's code-base.

Since the entire code-base is very large, we needed to visualize and comprehend each subsystem or architectural layer (as shown in Fig. 1) separately. The CodeCity model of the Applications layer, a total of 1902 classes, is pictured below in Fig. 5. Each block corresponds to a class. The width and height of each block denotes the number of class methods and attributes, respectively. The color intensity denotes the LOC of each class. Most of the larger structures are implementations (subclasses) of the class `Activities`

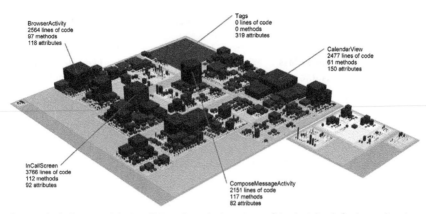

Fig. 5. CodeCity model visualizing the relative sizes of Android's default applications showing the imbalance in size and complexity between different classes.

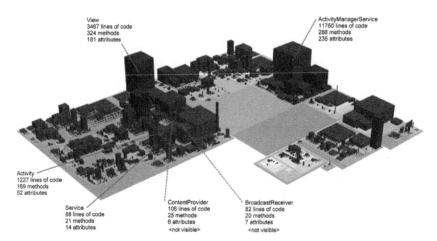

Fig. 6. CodeCity model of the Android application framework.

as described in Section 4.1. For example, `BrowserActivity` is the main user interface for the Browser application. `InCallScreen` is the interface shown to the user during a phone call.

Figure 6 depicts the CodeCity model generated from the Application Framework. A total of 3885 classes are contained within this model. A few selected classes, i.e., `Activity`, `Service`, `ContentProvider`, and `BroadcastReceiver` have been highlighted in Fig. 6. These are the classes which are extended in the Application layer when a developer wishes to implement one of the four major components of an Android application discussed in Section 4.1.

(a)

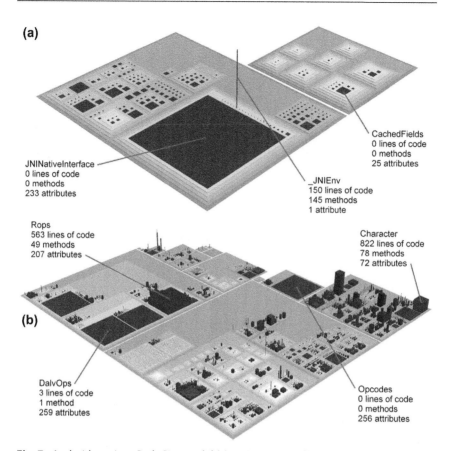

CachedFields
0 lines of code
0 methods
25 attributes

JNINativeInterface
0 lines of code
0 methods
233 attributes

_JNIEnv
150 lines of code
145 methods
1 attribute

Rops
563 lines of code
49 methods
207 attributes

Character
822 lines of code
78 methods
72 attributes

(b)

DalvOps
3 lines of code
1 method
259 attributes

Opcodes
0 lines of code
0 methods
256 attributes

Fig. 7. Android runtime CodeCity model (a): C/C++ source files, (b): Java source files.

The Android Runtime layer has two CodeCity models, one for its Java classes and one for its C/C++ files/classes—they can be seen in Fig. 7. The single thin standing file/class and large flat data file/class in the C/C++ model come from the implementation of the Java Native Interface. This interface uses a single object (JNIEnv) which is responsible for translating and delivering information back and forth between C/C++ and Java components within the system.

The Android Libraries layer, which contains 1036 classes/files, is displayed as a CodeCity model in Fig. 8. This model does not include all components of the Libraries layer, as the complete layer is much too large to model and render. A large portion of this layer is involved with graphics rendering: the Sk* classes indicated in Fig. 8 are a part of the Skia graphics library.

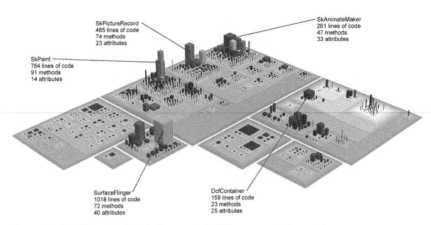

Fig. 8. Android libraries' CodeCity model (C/C++ source files only).

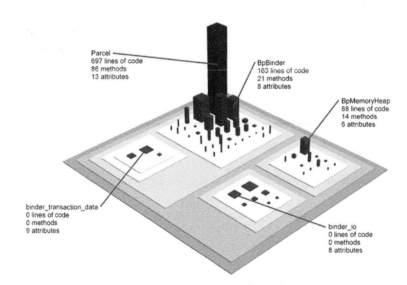

Fig. 9. Binder (IPC) CodeCity model (C/C++ source files only).

As well, `SurfaceFlinger` lives inside the Surface Manager, which is also responsible for graphics-based operations.

The Linux Kernel layer contains much less viewable code within the Android source; most of this layer comes from pre-built Linux Kernel files. Figure 9 shows the CodeCity model of the C/C++ Binder (IPC) component contained within this layer. The `Parcel` class (shown in Fig. 9) is used to package data during its transition between processes.

4.4 An Overview of the Automated Test Suites

List of Android's JUnit test suites, and brief statistics of the test suites (LOC, number of methods, and assertions) are presented in this section.

4.4.1 List of Test Suites

The test suites for the Android source code are developed by the Android team and are listed in an XML file located inside file: `development/testrunner/test_defs.xml`. This file defines 74 test suites. Not all of the suites were completely implemented and operational at the time this case study was conducted (Summer 2010). Table 4 shows the list of the test

Table 4 List of test suites for Android 2.1.

account	heap	cts-appsecurity
activity	imf	cts-content
android	improvider-permission	cts-database
apidemos	keystore-unit	cts-gesture
astl	launchperf	cts-graphics
browser	libcore	cts-hardware
browserfunc	libskia	cts-location
calendar	libstdcpp	cts-media
calprov	media	cts-net
camera	mediaapitest	cts-os
cameralatency	mediamemorystress	cts-perf1
camerastress	mediarecordertest	cts-perf2
contacts	mediastresstest	cts-perf3
contacts-launch	mediaunit	cts-perf4
gcontactsprov	mms	cts-perf5
contentprovider	mmslaunch	cts-permission
operation	musicplayer	cts-permission2
core	smoke	cts-provider
download	tablemerger	cts-telephony
provider-permission		
email	telephony-unit	cts-text
emailsmall	vpntests	cts-util
framework	cts-api-signature	cts-view
framework-permission	cts-api-signature-func	cts-webkit
globalsearch	cts-apidemos	cts-widget
globalsearch-permission	cts-app	cts-process

suites which were available including their names, whether they were operational and if we were able to measure their code coverage. Some of the test suites listed start with the acronym "cts," which stands for Compatibility Test Suite. The CTS is used by device manufacturers to ensure their product complies with the Android Compatibility Definition Document [27]. This document is used to ensure compatibility between Android devices and the applications which run on them.

Some of the test suites presented in the table do not provide code coverage functionality or are unable to execute due to various reasons. Several test suites are related to performance testing, and thus, do not provide code coverage such as cts-perf1 through cts-perf5 and launchperf. Some of the suites are written in C/C++ targeting C/C++ components of the platform such as libskia and libstdcpp. We experimented with several C/C++ code coverage tools in this platform (e.g., BullsEye C/C++ coverage tool [37]), however we were not successful in getting them to work on the Android base. The Emma code coverage tool proved effective for measuring code coverage for Java test suites.

Other tests are defined within test_defs.xml but have no implementation within the system. These include mmslaunch, browserfunc, and improvider-permission. These non-implemented test suites may still be defined even though their target code has been removed or perhaps their targets have yet to be implemented.

4.4.2 Statistics: LOC, Method, and Assertion Count

To gain an understanding of the size of the test suite, we measured its LOC, and the number of test methods and assertion calls in the test suite. Table 5 shows the LOC measures of the test suites which are generated by using Microsoft LOC Counter tool. Total test suite LOC is 376,955. Number of test files is 5,368, and total lines of comments are 209,201. The percentages

Table 5 LOC of Android 2.1 test suites.

		As a Percentage of Total Android Source (%)
C/C++	5624	0.09
Java	357,933	17.12
XML	12,610	1.99
Web files	788	0.14
Total	**376,955**	**4.03**

given in the rightmost column were calculated by comparing the LOC results here with those of Table 1 in Section 4.2.

As can be seen in Table 5, the C/C++ component of the test suites lags behind that of the Java component in number of methods, test methods, and assertions. In Section 4.2, it was estimated that a total of 179,553 test methods would have to be written in order to fully cover the 71,821 methods examined there. There are only 13,374 Java test methods in the test suites.

The results show that nearly all of the test code written for Android was done in Java at 357,933 LOC (95.0% of the total test suite LOC count). This suggests that a large portion of the Android code-base, the C/C++ component which makes up 63.4% of the entire Android source (Table 1), lacks direct testing by the main Android test suites. However, given that most of the C/C++ code-base is for the simplified Linux kernel and also the SQLite DBMS engine, and it is generally true that these two extra imported systems have been tested before by their own respective teams.

Approximated values for the method and assertion differences between Java and C/C++ and LOC differences between GUI and unit tests for the test suites were collected (see Table 6). As per Android SDK [1], unit tests in this platform are subclasses of class `AndroidTestCase`, and GUI tests are subclasses of `ActivityInstrumentationTestCase`. The counts of GUI and unit tests were derived by a code-base searching for those class names. An example of unit tests and a GUI test is shown in Figs. 10 and 11. It can be seen in these two examples that, although conceptually, each of these two test methods are one test case, they have several assert operations each. For this reason, we also counted the number of asserts in the test code-base. The average number of asserts per test method in JUnit code was about 2.5 (33,566/13,374).

Table 6 Various Android 2.1 test suite metrics (approximated).

Number of test methods per language	C/C++	32
	Java	13,374
	Total	13,406
Number of asserts per language	C/C++	772
	Java	33,566
	Total	34,338
LOC per test type	Unit Tests	213,834
	GUI Tests	71,179
Number of asserts per test type	Unit Tests	25,121
	GUI Tests	9,217

```
public void testProvider() throws Exception {
  long now = System.currentTimeMillis();

  Uri[] urls = new Uri[10];
  String[] dates = new String[]{
    Long.toString(new GregorianCalendar(1970, 1, 1, 0, 0, 0).getTimeInMillis()),
    Long.toString(new GregorianCalendar(1971, 2, 13, 16, 35, 3).getTimeInMillis()),
    Long.toString(new GregorianCalendar(1978, 10, 22, 0, 1, 0).getTimeInMillis()),
    Long.toString(new GregorianCalendar(1980, 1, 11, 10, 22, 30).getTimeInMillis()),
    Long.toString(now - (5 * 24 * 60 * 60 * 1000)),
    Long.toString(now - (2 * 24 * 60 * 60 * 1000)),
    Long.toString(now - (5 * 60 * 60 * 1000)),
    Long.toString(now - (30 * 60 * 1000)),
    Long.toString(now - (5 * 60 * 1000)),
    Long.toString(now)
  };

  ContentValues map = new ContentValues();
  map.put("address", "+15045551337");
  map.put("read", 0);

  ContentResolver contentResolver = mContext.getContentResolver();

  for (int i = 0; i < urls.length; i++) {
    map.put("body", "Test " + i + " !");
    map.put("date", dates[i]);
    urls[i] = contentResolver.insert(Sms.Inbox.CONTENT_URI, map);
    assertNotNull(urls[i]);
  }

  Cursor c = contentResolver.query(Sms.Inbox.CONTENT_URI, null, null, null, "date");

  for (Uri url : urls) {
    int count = contentResolver.delete(url, null, null);
    assertEquals(1, count);
  }
}
```

Fig. 10. An example unit test for SMS content provider.

```
@LargeTest
public void testArrowScrollDownToBottomElementOnScreen() {

  int numGroups = getActivity().getNumButtons();
  Button lastButton = getActivity().getButton(numGroups - 1);

  assertEquals("button needs to be at the very bottom of the layout for "
          + "this test to work",
          mLinearLayout.getHeight(), lastButton.getBottom());

  // move down to last button
  for (int i = 0; i < numGroups; i++) {
    sendKeys(KeyEvent.KEYCODE_DPAD_DOWN);
  }
  getInstrumentation().waitForIdleSync();
  assertTrue("last button should have focus", lastButton.hasFocus());

  int buttonLoc[] = {0, 0};
  lastButton.getLocationOnScreen(buttonLoc);
  int buttonBottom = buttonLoc[1] + lastButton.getHeight();
  assertEquals("button should be at very bottom of screen",
          mScreenBottom, buttonBottom);
}
```

Fig. 11. An example GUI test checking correct display of bottom fade on scrolling list of buttons.

5. RESULTS

RQs 1…5 are addressed in Sections 5.1–5.5, in order. Section 5.6 presents a summary of our results and their implications. Section 5.7 discusses the potential threats to the validity of our study and steps we have taken to minimize or mitigate them.

5.1 Code Coverage (RQ 1)

For RQ 1, we want to measure the extent of test coverage achieved by Android's test suites on various packages and components. This RQ has an exploratory rationale behind it, so as to explore the strength and completeness of Android's test suites.

5.1.1 Package-Level Coverage Measurement

After thorough examination, the test suites were executed and code coverage was collected using the coverage tool Emma [38], which is included in the Android package. The tool Emma provides measures for three types of coverage metrics: class coverage, method coverage, and line coverage.

Table 7 shows the class coverage, method coverage, and line coverage values. There are a total of 13 packages within Android which were targeted by the test suites for which coverage measurement is enabled by the Android team. The data for line and method coverage values are also visualized as two separate scatter plots in Fig. 12.

The total coverage values for all the targeted packages combined is fairly low at 41.4% of classes covered, 35.1% of methods covered, and 29.9% of lines covered. This perhaps means that extra efforts by Android's internal team and other contributors are needed to develop new test cases and increase test coverage. It also entails the major amount of work which has already been spent to achieve this relatively low coverage and the additionally needed major amount of work yet to be spent to increase coverage values.

According to both Table 7 and Fig. 12, it is obvious that the relative amount of effort for developing test cases for different packages is not the same. For example, considering line coverage, 63.72% of the lines of package GlobalSearch have been covered, while a very small ratio (only 0.05%) of package Calendar have been covered by the test suites. After careful review of the project context and similar studies (Section 2), we hypothesized the following reasons for such an imbalance in test strength and completeness across various packages: (1) implication of code reuse on testing and (2) risk-based testing [39], which are discussed next.

Table 7 Java code coverage values for test suite targets within the Android 2.1 source code.

Package	Classes			Methods			Lines		
	Covered	Total	Percent (%)	Covered	Total	Percent (%)	Covered	Total	Percent (%)
ApiDemos	10	509	1.96	26	1,873	1.39	89	7,933	1.13
Browser	4	187	2.14	11	1,189	0.93	64	6,784	0.94
Calendar	1	132	0.76	2	857	0.23	4	7,310	0.05
CalendarProvider	20	74	27.03	116	454	25.55	1,594	5,024	31.73
Camera	59	246	23.98	343	1,572	21.82	1,481	7,785	19.03
Contacts	23	197	11.68	135	1,342	10.06	657	8,798	7.47
DownloadProvider	0	24	0.00	0	131	0.00	0	1,759	0.00
Email	189	517	36.56	1,075	4,068	26.43	6,463	24,271	26.63
Framework	1,824	3,048	59.84	12,951	28,068	46.14	64,893	165,572	39.19
GlobalSearch	49	70	70.00	359	507	70.81	1,345	2,112	63.72
GoogleContacts Provider	3	92	3.26	10	974	1.03	341	9,782	3.49
Mms	72	307	23.45	372	2,402	15.49	1,477	11,436	12.92
Music	35	126	27.78	112	803	13.95	565	6,012	9.40
Total	2,289	5,529	41.40	15,512	44,240	35.06	78,977	264,578	29.85
Average			22.19			17.99			16.59

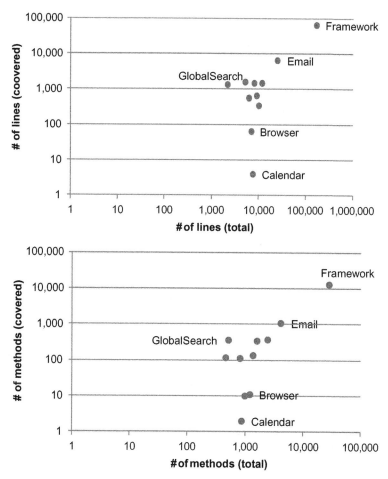

Fig. 12. Scatter plots for line and method coverage values.

In terms implication of code reuse on testing, it is expected that when a software package (e.g., the package `Calendar` in our context) is reused from another project, it does not have to be rigorously re-tested, if it has been tested well enough in its previous context. This might have been the case for package `Calendar` as it may have been reused from the Google's Calendar web service. However package `GlobalSearch` is specific to (has been developed for) Android, as it is intended to search all the phone's contents and perhaps the online databases. Thus, it does not seem to have been adapted from other Google web services.

For the potential relevance of risk-based testing [39] in this context impacting imbalance in test coverage values, it seems that the Android

development team may also have followed a risk-based heuristic for the
amount of testing (coverage) to be spent on each package. For example,
there may have been a reason based on which package `GlobalSearch`
was deemed to have more potential for defects, compared to for example,
package `Email`.

But note that both the above hypothesized reasons would need to be
analyzed and confirmed by proper analyses in future studies.

5.1.2 Fine-Grained (Class-Level) Coverage Measurement

In addition to conducting package-level coverage measurement (as reported
in the previous section), we also conducted more fine-grained coverage analysis in Java class-level. The scatter plot in Fig. 13 shows class-level coverage
value of each Java class versus its size (SLOC). Each point in this figure
corresponds to a Java class file. There are 2357 points (Java classes) in this
figure.

As we can observe, most of the points are positioned in the vertical
area to the left, indicating that most of the classes have relatively small
LOC sizes with varying coverage values. The largest Java class (in terms
of SLOC) is `TextView.java` inside package `android.widget` with
2,932 SLOC. There are also many classes with varying SLOC sizes with zero

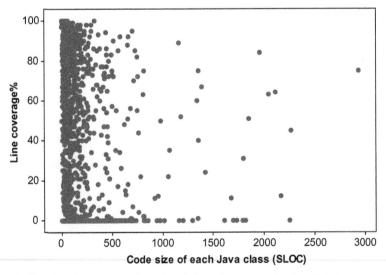

Fig. 13. Class-level coverage value of each Java class versus its size (SLOC).

or very low coverage values. There is no correlation between the two series as the Pearson correlation of coverage values and class size is only 0.01 (with P-value $= 0.337$).

5.1.3 Coverage Maps

Code coverage maps are 2-D visualizations to depict size of code versus the amount of coverage (e.g., [40] for the Linux test project). We used an online data visualization tool developed by IBM, called Many Eyes [41], for this purpose. The coverage maps for all the packages listed in Table 7 (except DownloadProvider) are shown and discussed next. Our goal for analyzing these coverage maps is to get a high-level view in terms of test coverage, for each package.

For each coverage map, the package under study has been broken down into its individual Java classes in order to show more detail. In the visualizations, the size of a box represents a particular class's Java SLOC. The larger the box, the larger the SLOC value) and the color represents its code coverage from white (no coverage) to a dark color (100% coverage).

We have conducted the analysis for all the 12 packages in Table 7. However, for brevity of this paper, we present the results and analysis for seven representative packages as shown in Fig. 14. Note that the names of the files (Java classes) inside the figures do not necessarily need to be readable in Fig. 14. Rather, it is important to analyze the general trend of the level of coverage by the heat-map aspects of these maps. As one can see, the extent of test coverage is not the same for all the classes inside each package and, thus, it is clear that the Android team has clearly spent different amount of focus on testing different classes. Two of these coverage maps (packages CalendarProvider and Camera) are shown in larger sizes in Figs. 15 and 16, and are discussed next. All the visualizations can also be accessed and dynamically analyzed online at [42].

The code coverage map for package CalendarProvider is shown in large size in Fig. 15. This package seems to have significantly more coverage than the Calendar package. CalendarProvider is used to access calendar-related information within the Android system. This provider is not only used by the Calendar application, but any third party application wishing to access the information, which may be why it has higher code coverage. Even for this package, the test coverage is more focused on the most critical Java classes as shown.

Figure 16 depicts the code coverage of the package Camera. In this package, the largest files, Camera.java and VideoCamera.java, both

Fig. 14. Code coverage map for seven selected Android packages. Metric: Line coverage.

of which are the main activities, a user sees when recording images, have a reasonable amount of code coverage (around 50%), however the package as a whole seems to be still lacking high amount of coverage.

As with other applications in Fig. 14, the coverage values for application (package) Contacts seem again relatively low. Also surprisingly, the main Contacts UI component (including classes ContactListActivity. java and ViewContactActivity.java) is not tested at all by the automated test suites.

For the Email package in Fig. 14, the left-hand side contains packages under com.android.email.* and com.android.exchange.* while the right-hand side of the visualization contains packages from org. apache.*, a third party library. As it can be seen, this third party library

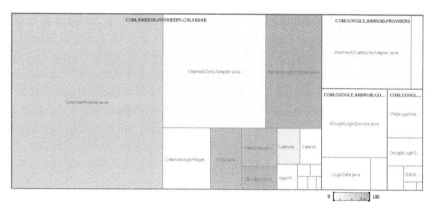

Fig. 15. Code coverage map of package `CalendarProvider`.

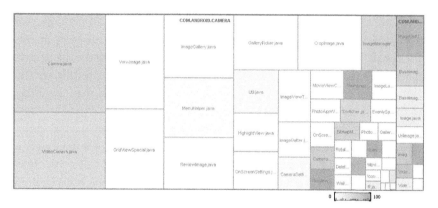

Fig. 16. Code coverage map of package Camera.

has some modest code coverage values, potentially revealing that Android developers wanted to perform their own testing to ensure its quality. After inspection, we found out that in fact the coverage of `org.apache` packages shown here are mostly due to indirect coverage, i.e., test cases are not explicitly targeting this area, but rather the calls to this package are made indirectly from other covered pieces of the code-base.

Much like the Email package, package `Framework` is split into two different sections. The left-hand side of the figure shows packages from `android.*` while the right contains `com.*` (mostly `com.android.internal.*`). By inspecting the packages presented in these two sections, a lot of redundancy was discovered. For example, while there is a package called `android.telephony`, there is also a package called `com.`

`android.internal.telephony`. Upon further investigation it is believed the reason for having two similar packages in different namespaces is one of security (we could not find any relevant documentation on Android website about this). The packages under `android.*` are open to the software development kit (SDK) for developers to use in their applications to access system functionality. The packages within `com.android.internal.*` are accessible by the Android system only and not open for use by application developers. Some of the packages within `android.*` rely on their `com.android.internal.*` counterparts in order to implement their functionality. As Table 7 showed, this package has the highest overall level of test coverage among all Android Java packages.

For the package `Mms`, while a few smaller files have a lot of coverage, the largest files are covered somewhat and much of this package lacks coverage. There are some packages which have no coverage at all.

In summary, after analyzing the test coverage results on all packages, we found that statement coverage seems to be more focused on components which will be used by application developers rather than the platform's default applications, such as Email and Browser. Exceptions to this are the `GlobalSearch` package, which is more of an application than anything, and the `ContactsProvider` package.

`GlobalSearch` is responsible for performing searches on the entire platform and the Internet is not directly used by application developers, yet it has the most relative coverage of the entire system. `ContactsProvider`, which would be used by application developers to access the contact information stored on the device, has relatively much less coverage. The Framework target has the highest statement coverage of the platform. This is most likely due to two reasons: (1) It is the largest package (in LOC) as it makes up most of the Application Framework layer from Section 4.1 and (2) It is most likely regarded as the most important package as it provides Android applications with access to the underlying libraries and hardware functionality of Android.

5.1.4 Coverage Histograms

In order to better visualize and analyze the coverage percentages in more depth, this section presents the histograms of coverage values for classes in a selected number of coverage targets (i.e., packages). The packages shown here were selected because they tend to have the widest range of coverage values.

Figure 17 shows the frequency of line coverage measure for different packages. As can be seen from these figures, a large number of classes have

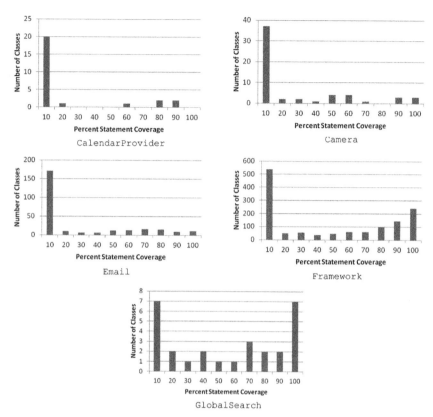

Fig. 17. Histogram of line coverage values for classes of different packages.

yet to be tested (a large number of classes have less than 10% coverage). While the Framework target does contain a large number of classes with 10% coverage or lower, it does have a fairly high number of classes with 71% coverage and above, breaking the pattern seen in the previous figures. The GlobalSearch target shows a remarkable change in the coverage patterns with just as many classes receiving 91% or higher coverage as those receiving 10% or lower.

Figure 18 below is a histogram showing the class frequency versus statement coverage percentages for all of the Java classes combined. Over 1,000 classes receive less than 10% coverage, but thanks to the Framework target, a large number of classes also reside at the other end of the spectrum. It is quite clear from this figure that the Android test suites could most likely need test enhancement as much of the system remains untested. This is even true when considering these coverage targets are only targeting Java components

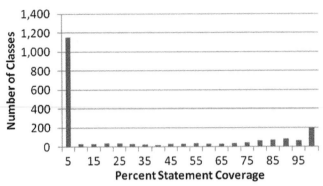

Fig. 18. Histogram of statement coverage values for all the Java classes in Android 2.1 code-base.

and neglect the C/C++ system components of which the test suites were found to be much smaller while the C/C++ component of the Android system is much larger.

5.2 Fault Detection Effectiveness of the Test Suites (RQ 2)

In both manual and automated testing, it is very important to measure fault detection effectiveness of test suites [20]. Mutation testing (i.e., controlled fault injection in the code-base) is the primary technique used for this purpose, in the software testing literature.

We provide next a brief overview of mutation testing first. We then discuss our attempt at selecting a mutation testing tool for the Android platform, and the scripts we have written to automate most of the mutation testing for this platform. We are hoping that, our script-set, would benefit other researchers and practitioner for conducting mutation testing on Android. We then present the results of mutation testing on Android.

5.2.1 An Overview of Mutation Testing

The goal of mutation testing is to systematically assess how effectively the available test suites of system detect common software faults [20]. The source code of the system under test (SUT) is systematically manipulated with each mutant copy receiving a single fault injected into it [20]. A mutant copy which does not compile is considered *stillborn* and not included in the rest of the process. The test suites are then run against each mutant. If the test suites detect the injected fault and produce a failure, the mutant is said to be *killed*. The test suites' *mutation score,*and perceived effectiveness, is the *ratio of killed mutants to total mutants generated.*

When generating mutants, it is always possible the injected defect may result in an *equivalent mutant*. Equivalent mutants are ones which, for all intents and purposes, have the same functionality as the SUT [20]. While some researchers suggest that equivalent mutants should be removed from the study before calculating the mutation score, some other researchers have suggested simply ignoring equivalent mutants [43] as it is very difficult to detect them with an automated process.

5.2.2 Selecting a Mutation Testing Tool for the Android Platform

The following six different mutation tools were evaluated for potential use in this study: muJava [44], Lava [45], Jumble [46], Judy [47], Javalanche [48], and Jester [49]. Technological challenges make working with mutation tools on ultra-large code-bases such as Android (9 million LOC) non-trivial. For example, it was found that muJava, Jumble, and Javalanche all try to mutate compiled Java byte-code files instead of the source code files themselves. This proved unsuitable for a mutation study on Android due to the complicated nature of its build process. Additionally, the required directory tree structure of the muJava tool is highly incompatible with the Android code-base. The Judy tool was found to be inoperable on Linux, which is required to build Android, as it was looking for user files in a Windows directory. The Jester tool requires that the test suites do not fail on the non-mutated version of the SUT, which proved not to be the case with Android 2.1 running on the Android emulator since some tests failed under this circumstance.

Lava was chosen for this project because of its simplicity and less automated behavior. Lava mutates a single target source file and provides a copy without attempting to access byte code or run the tests itself. This allowed for the production of custom build scripts to handle the building, running, and testing of the mutants while Lava simply provided mutants.

Two major issues encountered while using the Lava tool were circumvented by modifying the custom build scripts: (1) One issue was Lava's inability to work with Java files that use generics and annotations. This issue was fixed by replacing generics and annotations with placeholders until after the mutation occurred. (2) Because Lava is a simple tool, it has no memory of previous mutations which means it is entirely possible to create the same mutant twice. This issue was resolved by checking the current mutant with all previous mutants.

5.2.3 Mutation Operators

One disadvantage with Lava is the limited number of mutation operators available with this simple mutation tool. As can be seen from Table 8, all the

Table 8 The mutation operators provided with the Lava tool [45] are non-object-oriented with no class-level object-oriented mutation operators available.

Mutation Operator	Mathematical/Logical Operators That are Affected
Boolean constants	true, false
Boolean operators	&&, ‖
Relational operators	$==, <=, >=, <, >, !=$
Increment/Decrement operators	$++, --$
Arithmetic operators	$+, -, *, /$
Binary bit operators	$\&, |, \char`\^, \%, \ll, \gg, \ggg$
Arithmetic assignment operators	$+=, -=, *=, /=$
Binary bit assignment operators	$\&=, |=, \char`\^=, \%=, \ll=, \gg=, \ggg=$

mutation operators provided by Lava are non–object-oriented [20]; there are no class-level object-oriented mutation operators. Furthermore, Lava only replaces a mathematical/logical operators in one category by an operator in the same category (i.e., "==" can only be replaced by "<=," ">=," "<," ">" or "! =" and with &, | ,‖ etc.).

5.2.4 Our Mutation Script-Set

We conducted extensive work to develop custom scripts to run Android's test suites and our custom-made mutations testing platform. All the custom scripts that we developed have been made available online at [50] for replicability purposes and further usage by other researchers and practitioners. To use and work with these scripts, guidelines have also been provided online.

5.2.5 Mutation Testing in Android

Using the custom-made mutation testing platform [50] that we developed, we conducted mutation analysis on the platform. Conducting full-scale mutation testing on Android was impossible due to major computational challenges (due to nature of mutation testing) as discussed below. We thus selected a representative Java class inside each package for mutation and conducted mutation testing on them only, as shown in Table 9.

To see how the above values have been calculated, let us discuss the file `Activity.java` in folder `frameworks/base/core/java/android/app/` as an example mutation target. This file defines the `android.app.Activity` class and as such represents an important component in the Android system. The results of the mutation testing of this class can be seen in Table 10.

Table 9 Results from mutation testing of several Java classes.

Package	Representative Java Class Selected for Mutation	Mutation Score (%)
Browser	`com.android.browser/BrowserActivity.java`	2.4
Calendar	`com.android.calendar/AlertActivity.java`	1.0
CalendarProvider	`com.android.providers.calendar/CalendarProvider.java`	35.4
Camera	`com.android.camera/Camera.java`	25.3
Contacts	`com.android.contacts/ContactsUtils.java`	23.7
ContactsProvider	`com.android.providers.contacts/ContactsProvider.java`	15.2
Email	`com.android.email/Account.java`	45.5
Mms	`com.google.android.mms/ContentType.java`	24.6
Music	`com.android.music/MediaPlaybackActivity.java`	25.4
Framework	`frameworks/base/core/java/android/app/Activty.java`	71.2
DownloadProvider	`com.android.providers.downloads/DownloadProvider.java`	0.0
GlobalSearch	`com.android.providers.contacts/GlobalSearchSupport.java`	50.3

Table 10 Results from mutation testing of Java class `Activty.java`.

Total mutants created	85
Equivalent mutants	5
Stillborn mutants	28
Total mutants tested	52
Killed mutants	37
Mutation score	71.2%

To provide a snapshot on the major time complexity of mutation testing, as shown in Table 10, the Lava tool produced 85 mutants for the class `Activty.java`. Using our custom-made mutation testing platform, running the entire Android test suites on all the 85 mutants took about 31 h (about 21.88 min for each mutant) on a conventional PC. Our experiment thus emphasizes again the major computational challenge of applying mutation testing on large systems and its scalability limitations [51].

5.3 Code Coverage Versus Number of Reported Actual Defects (RQ 3)

For RQ 3, we wanted to assess whether code coverage values correlate with the number of actual reported defects. We would expect that, for packages with larger coverage values (meaning more rigorous testing), less defects would be reported by users after release. Also, on the contrary, is it necessarily true that components with low coverage have more defects? Or from another perspective, do the development/testing teams tend to develop better test suites with more coverage for components with more defects?

To answer RQ 3, we already had calculated the code coverage of various packages in Section 5.1. To measure the number of reported actual defects for each package, we mined the Android bug repository by proving each of the package names as the search keyword and ensuring that the issue type is "defect." An example search screen is shown in Fig. 19, in which the search is done for the number of reported actual defects for package Camera, and the output value is 776.

Based on the two data series (code coverage values and the actual number of defects), we visualize the data as a scatter plot in Fig. 20.

As we can observe, in contrary to what one would expect, for packages with larger coverage values (meaning more rigorous testing), it is not necessarily true that less defects have been reported by the users after release. Also, it is not necessarily true that components with low coverage have more defects.

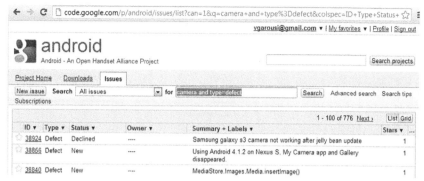

Fig. 19. Mining the Android bug repository for the number of reported actual defects for each package.

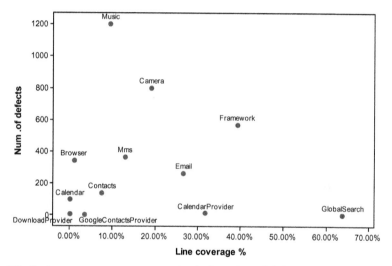

Fig. 20. Code coverage versus reported actual number of defects.

To see whether there was any correlation between code coverage and number of reported defects, we calculated the Pearson correlation of the x and y values in Fig. 20. The correlation value is 0.06 and P-value is 0.83, indicating the absence of any correlation.

5.4 Correlation Between Code Coverage and Fault Detection Effectiveness (RQ 3)

By putting together the data for code coverage (Section 5.1) and mutation score (Section 5.2), we present the scatter plot in Fig. 21. To see whether there was any correlation between code coverage and mutation score, we calculated

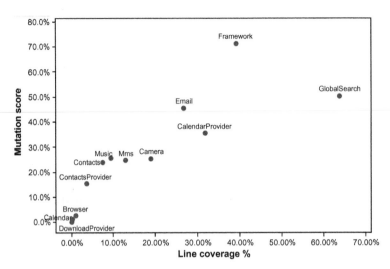

Fig. 21. Code coverage versus mutation score.

the Pearson correlation of the x and y values in Fig. 21. The correlation value is 0.84 and P-value is 0.001, indicating the existence of a quite strong correlation. Existence of correlation between code coverage and mutation score has also been shown by evidence in other studies as well, e.g., [52].

5.5 Defect Density (RQ 5)

Recall that defect density is defined as the average number of defects per thousand lines of code. By putting together the data for LOC measures (Section 4.2) and number of reported defects (Section 5.3), we present the scatter plot in Fig. 22. The package with the highest defect density (DD) is Music (DD = 0.19) and the package with the lowest DD value is ContactsProvider (DD = 0.0003).

To see whether there was any correlation between LOC size and number of reported defects, we calculated the Pearson correlation of the x and y values in Fig. 22. The correlation value is 0.10 and P-value is 0.76, indicating the existence of a very week correlation.

It would be logical to compare our DD measures with those reported by Coverity Inc. [53], where the firm found 359 defects in the Android kernel that is used in the phone brand *HTC Droid Incredible*. That report's calculations showed that the Android kernel defect density is 0.47 defects per 1000 lines of code, being better than the industry average on one defect per 1000 lines of code. The highest defect density value in our analysis was

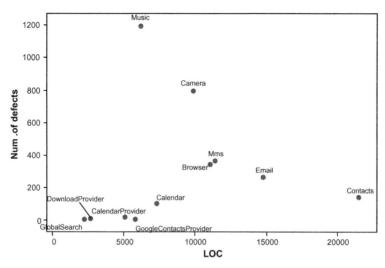

Fig. 22. Defect density.

0.19 (per 1000 lines of code), lower than the Android kernel defect density (0.47) reported in [53].

5.6 Summary of Results and Implications

Our RQs 1…5 addressed various aspects of Android test suite. A summary of all the results and their implications are discussed below:

RQ 1:

- *Summary:* Test coverage measurements (Section 5.1): The total coverage values for all the targeted packages combined are fairly low: 41.4% of classes were covered, 35.1% of methods, and 29.9% of lines were covered. As a summary of measures reported in Section 5.1, Fig. 23 shows the individual-value plot of coverage values for different packages of Android for four different coverage metrics. We also noticed that the variances of the coverage values for different packages are quite high, indicating the difference in amount of testing efforts spent on different packages.
- *Implications:* Results of RQ 1 demonstrated the variance in coverage measures across difference packages and difference classes of each package. As implications, those results call for further research and investigation into the root cause of spending varying amount of test efforts on different parts of the system and whether any of the following two example factors have been the case in the context of Android test suite development: (1) implication of code reuse on testing, and (2) risk-based testing [39].

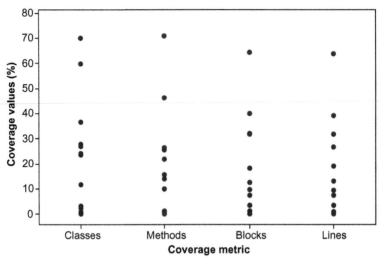

Fig. 23. Summary of measures reported in Section 5.1: Individual-value plots of coverage values for different packages of Android for four different coverage metrics. Each point corresponds to a package, e.g., Camera.

RQ 2:

- *Summary:* Analysis of fault detection effectiveness using mutation testing (Section 5.2): Similar to coverage values, mutation testing scores also varied for selected classes of different packages. Also, our experiment emphasized again the major computational challenge of applying mutation testing on large systems and its scalability limitations [51]. For instance, to provide a snapshot on the major time complexity of mutation testing in our context, the Lava mutation tool produced 85 mutants for the class `Activty.java`. Using our custom-made mutation testing platform, running the entire Android test suites on all those 85 mutants took about 31 h (about 21.88 min for each mutant) on a conventional PC.
- *Implications:* Mutation testing results of RQ 2 imply the need for further work by researchers and practitioners into improving the power (fault detection effectiveness) of Android test suites.

RQ 3:

- *Summary:* Code coverage versus number of reported actual defects (Section 5.3): As we observed and analyzed, in contrary to what one would expect, for packages with larger coverage values (meaning more rigorous testing), it is not necessarily true that less defects have been reported by the users after release. Also, it is not necessarily true that components with low coverage have more defects.

- *Implications:* Results of RQ 3 imply the need for further research and investigation into the root causes of spending varying amount of test effort across different components.

RQ 4:

- *Summary:* Cost-effectiveness of code coverage, i.e., correlation between code coverage and fault detection effectiveness (Section 5.4): We re-confirmed (replicated [28]) the existence of correlation between code coverage and mutation score, similar to other studies as well, e.g., [52].
- *Implications:* Replicated (re-confirmed) results from RQ 4 (for correlation between code coverage and mutation score) are valuable in the area of evidence-based software testing [28].

RQ 5:

- *Summary:* Defect density of different packages (Section 5.5): We assessed how size metrics (LOC) of different packages correlated with the number of defects reported for them. The package with the highest defect density (DD) was `Music` (DD = 0.19 per 1 KLOC) and the package with the lowest DD value is `ContactsProvider` (DD = 0.0003). Also we found that the highest defect density value in our analysis (0.19 per 1000 lines of code) was lower than the Android kernel defect density (0.47) reported in [53].
- *Implications:* Similar to other studies, e.g., [53], defect density measures can be used in future studies to investigate the root cases (e.g., code complexity) of higher defect density for certain packages/classes compared to others.

5.7 Threats to Validity and Their Mitigation

Potential threats to the validity of our study and steps we have taken to minimize or mitigate them are discussed below:

External validity: The test cost-effectiveness and defect density analysis was conducted on one real large-scale software, i.e., Android. Effort was taken to conduct the study in the most realistic approach possible, however similar to other case studies, generalizing results to other systems is not easily possible.

Construct validity: Construct validity is the extent to which what was to be measured was actually measured. We used the well-known metrics for source code size, complexity, test coverage, and mutation testing.

Conclusion validity: The conclusions and implications that we derived from our results (Section 5.1–5.6) were based on actual quantitative numbers, without researchers' bias.

6. CONCLUSIONS AND FUTURE WORKS

The Android operating system is one of the most popular open-source platforms in the mobile operating system market. It had a worldwide smartphone market share of 68% at the second quarter of 2012. However, there has been little research on test coverage and test cost-effectiveness in this platform. The goal of this case study reported in this paper was to assess test coverage, fault detection effectiveness, test cost-effectiveness, and defect density in code-base of version 2.1 of the Android platform. We raised and addressed five research questions (RQs) in this study. Among our results were: (1) in contrary to what one would expect, for packages with larger coverage values (meaning more rigorous testing), it is not necessarily true that less defects have been reported by the users after release. Also, it is not necessarily true that components with low coverage have more defects; (2) we re-confirm (replicate) the existence of correlation between code coverage and mutation score, similar to existing studies; and (3) the package with the highest defect density (DD) in the Android code-base is Music (DD = 0.19 per 1 KLOC) and the package with the lowest DD value is ContactsProvider (DD = 0.0003). Results of our study have helped us and it is hoped that they will help other researchers to get a better view on test coverage, fault detection effectiveness, test cost-effectiveness, and defect density in Android code-base.

One future work direction is to replicate this study on newer versions of Android and assess the changes in trends for the metrics that were measured.

ACKNOWLEDGMENTS

Vahid Garousi was supported by the Discovery Grant No. 341511–07 from the Natural Sciences and Engineering Research Council of Canada (NSERC) and by the Visiting Scientist Fellowship Program (#2221) of the Scientific and Technological Research Council of Turkey (TUBITAK). Riley Kotchorek was supported by the NSERC through the Undergraduate Student Research Awards Program (USRA). Michael Smith was supported by the Discovery Grant No. 1754–2009 from the Natural Sciences and Engineering Research Council of Canada (NSERC). Michael Smith and Vahid Garousi were also supported through an NSERC industrial collaborative research and development grant CRDPJ/365295–2008 with Analog Devices and CDL Systems.

REFERENCES

[1] Android Source Code Base. <http://source.android.com>(retrieved October 2010).

[2] Members of the Open Handset Alliance. <http://www.openhandsetalliance.com/oha_members.html> (retrieved January 2011).

[3] International Data Corporation (IDC). Android and iOS surge to new Smartphone OS record in second quarter, according to IDC. <http://www.idc.com/getdoc.jsp?containerId=prUS23638712> (retrieved October 2012).

[4] Coverity Inc., Coverity Scan: 2010 Open Source Integrity Report-Android Kernel, 2010. <http://www.coverity.com//coverity-scan-2010-open-source-integrity-report.pdf> (retrieved 24.10.12).

[5] Coverity Inc., A look inside the Android Kernel with automated code testing, 2011. <http://www.coverity.com/library/pdf/a-look-inside-the-android-kernel.pdf> (retrieved 24.10.12).

[6] A. Kumar Maji, K. Hao, S. Sultana, S. Bagchi, Characterizing failures in mobile OSes: a case study with Android and Symbian, in: International Symposium on Software Reliability Engineering, 2010, pp. 249–258.

[7] M. Asaduzzaman, M.C. Bullock, C.K. Roy, K.A. Schneider, Bug introducing changes: a case study with Android, in IEEE Working Conference on Mining Software Repositories, 2012, pp. 116–119.

[8] L. Martie, V.K. Palepu, H. Sajnani, C.V. Lopes, Trendy bugs: topic trends in the Android bug reports, in: IEEE Working Conference on Mining Software Repositories, 2012, pp. 120–123.

[9] V. Guana, F. Rocha, A. Hindle, E. Stroulia, Do the stars align? Multidimensional analysis of Android's layered architecture, in: IEEE Working Conference on Mining Software Repositories, 2012, pp. 124–127.

[10] W. Hu, D. Han, A. Hindle, K. Wong, The build dependency perspective of Android's concrete architecture, in: IEEE Working Conference on Mining Software Repositories, 2012, pp. 128–131.

[11] V.S. Sinha, S. Mani, M. Gupta, MINCE: mining change history of Android project, in: IEEE Working Conference on Mining Software Repositories, 2012, pp. 132–135.

[12] L.A. Reina, G. Robles, Mining for localization in Android, in: IEEE Working Conference on Mining Software Repositories, 2012, pp. 136–139.

[13] M. Hutchins, H. Foster, T. Goradia, T. Ostrand, Experiments of the effectiveness of data flow- and control flow-based test adequacy criteria, in: Proceedings of the International Conference on Software Engineering, 1994, pp. 191–200.

[14] J.H. Andrews, L.C. Briand, Y. Labiche, A.S. Namin, Using mutation analysis for assessing and comparing testing coverage criteria, IEEE Trans. Softw. Eng. 32 (8) (2006) 608–624.

[15] P.G. Frankl, S.N. Weiss, C. Hu, All-uses vs mutation testing: an experimental comparison of effectiveness, J. Syst. Softw. 38 (3) (1997) 235–253.

[16] A. Gupta, P. Jalote, An approach for experimentally evaluating effectiveness and efficiency of coverage criteria for software testing, Int. J. Softw. Tools Technol. Transf. 10 (2) (2008) 145–160.

[17] S.A. Jolly, V. Garousi, M.M. Eskandar, Automated unit testing of a SCADA control software: an industrial case study based on action research, in: Proceedings of the IEEE International Conference on Software Testing, Verification and Validation, 2012, pp. 400–409.

[18] K.I. Seo, E.M. Choi, Comparison of five black-box testing methods for object-oriented software, in: Proceedings of the International Conference on Software Engineering Research, Management and Applications, 2006.

[19] P.G. Frankl, O. Iakounenko, Further empirical studies of test effectiveness, ACM SIGSOFT Softw. Eng. Notes 23 (6) (1998) 153–162.

[20] A.P. Mathur, Foundations of Software Testing, Addison-Wesley Professional, 2008.

[21] Software Reliability LLC, Current defect density statistics. <http://www.softrel.com/Current%20defect%20density%20statistics.pdf> (retrieved 24.10.12).

[22] S. McConnell, Code Complete: A Practical Handbook of Software Construction, second ed., Microsoft Press, 2004.

[23] H. Kagdi, M.L. Collard, J.I. Maletic, A survey and taxonomy of approaches for mining software repositories in the context of software evolution, J. Softw. Maint. Evol.: R. 19 (2) (2007) 77–131.

[24] V. Garousi, Evidence-based insights about issue management processes: an exploratory study, in: Proceedings of the International Conference on Software Process (ICSP), 2009, pp. 112–123.

[25] A. Zaidman, B. Van Rompaey, S. Demeyer, A. van Deursen, Mining software repositories to study co-evolution of production & test code, in: International Conference on Software Testing, Verification, and Validation, 2008, pp. 220–229.

[26] G. Barahona, G. Robles, Applying social network analysis to the information in CVS repositories, in: Proceedings of the Mining Software Repositories Workshop, 2004.

[27] O.S. Gómez, N. Juristo, S. Vegas, Replication, reproduction and re-analysis: three ways for verifying experimental findings, in: International Workshop on Replication in Empirical Software Engineering Research, 2010.

[28] F.J. Shull, J. Carver, S. Vegas, N. Juristo, The role of replications in empirical software enginering, Empir. Softw. Eng. 13 (2) (2008) 211–218.

[29] P. Runeson, H. Martin, Guidelines for conducting and reporting case study research in software engineering, Empir. Softw. Eng. 14 (2) (2009) 131–164.

[30] V. Basili, G. Caldiera, D.H. Rombach, The Goal Question Metric approach, in: J. Marciniak (Ed.), Encyclopedia of Software Engineering, Wiley, 1994.

[31] Android Open Source Project. <http://android.git.kernel.org/> (retrieved October 2010).

[32] Microsoft Corp., Microsoft LOC counter tool. <http://archive.msdn.microsoft.com/LOCCounter> (retrieved October 2012).

[33] G.K. Gill, C.F. Kemerer, Cyclomatic complexity density and software maintenance productivity, IEEE Trans. Softw. Eng. 17 (12) (1991) 1284–1288.

[34] F. Sauer, Eclipse metrics plug-in. <http://metrics.sourceforge.net/> (retrieved October 2010).

[35] S. Diehl, Software Visualization: Visualizing the Structure, Behaviour, and Evolution of Software, Springer, 2007.

[36] R. Wettel, CodeCity tool. <http://codecity.inf.usi.ch> (retrieved October 2010).

[37] Bullseye Testing Technology, BullsEye C/C++ coverage tool. <www.bullseye.com> (retrieved October 2012).

[38] EMMA: a free Java code coverage tool. <http://emma.sourceforge.net> (retrieved October 2010).

[39] S. Biffl, A. Aurum, B. Boehm, H. Erdogmus, P. Grünbacher, Value-Based Software Engineering, Springer, 2005.

[40] Linux Test Project, Coverage galaxy map. <http://ltp.sourceforge.net/coveragemap.php> (retrieved October 2010).

[41] IBM, About many eyes. <http://manyeyes.alphaworks.ibm.com/manyeyes/page/about.html> (retrieved October 2010).

[42] R. Kash, V. Garousi, Code coverage maps (Visualizations) for Android test suites. <http://www-958.ibm.com/software/data/cognos/manyeyes/users/RileyKash/visualizations> (retrieved January 2011).

[43] A.J. Offutt, R.H. Untch, Mutation 2000: uniting the orthogonal, in: Mutation Testing for the New Century, Kluwer Academic Publishers, 2001, pp. 34–44.

[44] J. Offutt, MuJava Home Page (JMutation), from George Mason University Department of Computer Science: <http://cs.gmu.edu/~offutt/mujava> (retrieved August 2010).

[45] S. Danicic, Lava: a system for mutation testing of Java programs, 2010. <http://www.doc.gold.ac.uk/~mas01sd/mutants/index.html> (retrieved August 2009).

[46] Jumble, 2010. <http://jumble.sourceforge.net/> (retrieved August 2009).

[47] Projects – Judy, Software Engineering Society, 2010. <http://www.e-informatyka.pl/sens/wiki.jsp?page=projects.judy> (retrieved August 2009).

[48] Javalanche – About, Saarland University, 2010. <http://www.st.cs.uni-saarland.de/~schuler/javalanche/index.html> (retrieved August 2009).

[49] Jester, SourceForge, 2010. <http://jester.sourceforge.net> (retrieved August 2009).

[50] R. Kash, V. Garousi, Scripts to run Android's Tests and Mutations. <http://www.softqual.ucalgary.ca/projects/2011/Android/Scripts_to_run_Android_Tests_and_Mutations.zip> (retrieved January 2011).

[51] Y. Jia, M. Harman, An analysis and survey of the development of mutation testing, IEEE Trans. Softw. Eng. 37 (5) (2011) 649–678.

[52] S.A. Jolly, V. Garousi, M.M. Eskandar, Automated unit testing of a SCADA control software: an industrial case study based on action research, in: IEEE International Conference on Software Testing, Verification and Validation (ICST), 2012, pp. 400–409.

[53] Coverity Inc., Coverity scan: 2010 open source integrity report-Android Kernel. <http://www.coverity.com/library/pdf/coverity-scan-2010-open-source-integrity-report.pdf> (retrieved October 24 2012).

ABOUT THE AUTHORS

Vahid Garousi is a Visiting Professor of Software Engineering in the Middle East Technical University (METU) in Ankara, Turkey. He is also an Associate Professor at the University of Calgary, Canada. He is also a practicing software engineer and coach, and provides consultancy and corporate training services in the North America in the areas of software testing and quality assurance, model-driven development, and software maintenance. During his career, Vahid has been active in initiating several major software testing and software engineering projects in Canada with the Canadian and multi-national software companies. Vahid completed his PhD in Software Engineering in Carleton University, Canada, in 2006. His PhD work was on performance testing of real-time software systems. He has been involved in different software engineering conference committees as an organizing or program committee member, such as ICST, ICSP, CSEE&T, and MoDELS. He is a member of the IEEE and the IEEE Computer Society, and is a licensed professional engineer (PEng) in the Canadian province of Alberta. He has been selected as a Distinguished Visitor (speaker) for the IEEE Computer Society's Distinguished Visitors Program (DVP) for the period of 2012–2014. Among his awards is the Alberta Ingenuity New Faculty Award in June 2007. His research interests include software engineering, software testing and quality assurance, model-driven development, and software maintenance.

Riley Kotchorek has a B. Sc. in Computer Engineering from the University of Calgary. He graduated at the top of his class in 2011. After graduation, he worked for an engineering company providing SCADA devices for energy utilities. He now works within the oil industry in Calgary. His interests include programming entertainment software and computer networking.

Michael Smith obtained his B. Sc. in physics from the University of Hull, UK and his PhD in physics from the University of Alberta, Canada. After a number of years teaching science and mathematics in secondary schools, he returned to academia as a professor in electrical and computer engineering at the University of Calgary, Canada. In 1994, he received the Canadian National Wighton Fellowship from the Sandford Fleming Foundation for his work on Innovative teaching within Undergraduate Engineering Laboratories. In 2003, he became an adjunct professor with the Department of Radiology, University of Calgary, Canada. In 2012, he received a Killam Trust Interdisciplinary Research Award for his collaborative bio-medical engineering research and was again recognized for his approach to undergraduate teaching. His major research interests involve the application of software engineering and customized real-time digital signal processing algorithms in the context of mobile-embedded systems and bio-medical instrumentation.

AUTHOR INDEX

SUBJECT INDEX

computing configuration change
impact, 156
computing configuration
differences, 156
configuration aware testing, 143
configuration definition layer
(CDL), 143
coverage- and fault detection-based
weighting, 158
interaction benefit-based
prioritization, 157
prioritization, 157
regression testing, 145, 148
sampling or selection technique, 144
selecting test cases, 157
selection, 150
single version, 147

solutions and techniques, 149
specification-based
weighting, 159
strategies, 146
TouchUtils, 20

U
UI/Application Exerciser
Monkey, 45
Unit testing
Activity testing, 18
Service testing, 21
Content Provider testing, 22
Usability testing, 42

V
vim, 146

CONTENTS OF VOLUMES IN THIS SERIES

Volume 71

Volume 72

Volume 73

Volume 87

Introduction and Preface
    SAHRA SEDIGH AND ALI HURSON
Techniques to Measure, Model, and Manage Power
    BHAVISHYA GOEL, SALLY A. MCKEE, AND MAGNUS SJÄLANDER
Quantifying IT Energy Efficiency
    FLORIAN NIEDERMEIER, GERGŐ LOVÁSZ, AND HERMANN DE MEER
State of the Art on Technology and Practices for Improving the Energy Efficiency of Data
Storage
    MARCOS DIAS DE ASSUNÇÃO AND LAURENT LEFÈVRE
Optical Interconnects for Green Computers and Data Centers
    SHINJI TSUJI AND TAKASHI TAKEMOTO
Energy Harvesting for Sustainable Smart Spaces
    NGA DANG, ELAHEH BOZORGZADEH, AND NALINI VENKATASUBRAMANIAN

Volume 88

Energy-Aware High Performance Computing—A Survey
    MICHAEL KNOBLOCH
Micro-Fluidic Cooling for Stacked 3D-ICs: Fundamentals, Modeling and Design
    BING SHI AND ANKUR SRIVASTAVA
Sustainable DVFS-Enabled Multi-Core Architectures with On-Chip Wireless Links
    JACOB MURRAY, TENG LU, PARTHA PANDE, AND BEHROOZ SHIRAZI
Smart Grid Considerations: Energy Efficiency vs. Security
    ANDREAS BERL, MICHAEL NIEDERMEIER, AND HERMANN DE MEER
Energy Efficiency Optimization of Application Software
    KAY GROSSKOP AND JOOST VISSER

Printed and bound by CPI Group (UK) Ltd, Croydon, CR0 4YY

03/10/2024

01040425-0001